MAKE 50
WILD AND WACKY
(but Useful!)
CONTRAPTIONS

MAKE 50 WILD AND WACKY
(BUT USEFUL!) CONTRAPTIONS

HarperCollins books may be purchased for educational, business,
or sales promotional use. For information please write:
Special Markets Department, HarperCollins Publishers,
10 East 53rd Street, New York, NY 10022

FIRST EDITION

The name of the "Smithsonian," "Smithsonian Institution," and the sunburst
logo are registered trademarks of the Smithsonian Institution.

Rube Goldberg is the ® and © of Rube Goldberg, Inc.
www.RubeGoldberg.com

Image on p. 6 (Goldberg, Ruben—POR 4) courtesy of the Bancroft Library,
University of California, Berkeley

Conceived, designed, and produced by Quid Publishing,
Level 4 Sheridan House, 114 Western Road, Hove, BN3 1DD
United Kingdom
www.quidpublishing.com

Design: Lindsey Johns
Illustrations: Robert Brandt

Library of Congress Cataloging-in-Publication Data

Chaline, Eric.
  Make 50 wild and wacky (but useful!) contraptions / Eric Chaline, Robert
Brandt. — 1st ed.
      p. cm.
  ISBN 978-0-06-143776-2
1. Inventions—Handbooks, manuals, etc. I. Brandt, Robert, 1970- II. Title.
III. Title: Wild and wacky but useful contraptions. IV. Title: Make fifty wild
and wacky (but useful!) contraptions.

T339.C45 2008
620--dc22

                                                                2007032283

MAKE 50
WILD AND WACKY
(but Useful!)
CONTRAPTIONS

Eric Chaline and Robert Brandt

Collins
An Imprint of HarperCollinsPublishers

CONTENTS

Introduction

Welcome to *Make 50 Wild and Wacky (but Useful!) Contraptions*. Before we get started on the projects themselves, let's investigate some of the individuals who provided contraptions with a long and rich history.

True to its title, this book tells you all you need to know about designing and building "contraptions." In common parlance, a contraption is a machine that is overly complicated for its purpose; it might also be one that isn't particularly well made, and by implication, one that is not very safe. Our own reading of the word has much more to do with the former meaning (stripped of its negative connotations) than the latter. Another name for such contraptions is "Rube Goldberg machines," which *Webster's Dictionary* defines as: "Machines accomplishing by extremely complex roundabout means what actually or seemingly could be done simply." There you have it in a nutshell, but what the dictionary omits are the crucial elements of creativity, fun, surprise, and satisfaction that are such an important part of "contraption engineering."

Rube Goldberg became so famous for his cartoon machines that they were later named after him.

The Man Behind the Machines

The Rube Goldberg (1883–1970) for whom the machines are named was a native of San Francisco. A graduate in engineering from the University of California, his first, and unlikely, job was as an engineer in the City of San Francisco Water and Sewers Department. Luckily for the world, his true vocation of drawing cartoons quickly asserted itself. After six months he quit his job as a city engineer and joined the staff of the *San Francisco Chronicle*, which soon published his first cartoons. In 1907, he moved to New York to work on the *Evening Mail*. His cartoons were syndicated from 1915, and he quickly became one of the US's most famous and best-loved cartoonists. He went on to found the National Cartoonist Society, and won the Pulitzer Prize for his political cartoons in 1948; but it's for his drawings of "absurdly connected machines"—soon to be rechristened Rube Goldberg machines in his honor—that he's best remembered.

Although Rube Goldberg won the greatest fame for his absurd mechanical creations, he wasn't the only cartoonist of the period to hit upon the idea of overly complex machines performing simple tasks. Two other men, like Rube born in the closing decades of the nineteenth century, shared his passion for cartoons and fantastic engineering: the Englishman Heath Robinson (1872–1944); and the Dane Robert Storm Petersen (1882–1949). It's intriguing to think that these three men, from such different countries and backgrounds, came up with such similar ideas and presented them in the same medium of cartoons. It's always possible that one influenced the other two, but another explanation suggests itself. The three grew up during the dawning of the technological age when the automobile, the telephone, and electricity were revolutionizing everyday life. Their drawings poke fun at those who embraced this new world of technology too enthusiastically, preferring some overly complex mechanical monstrosity to the simpler human way of doing things. Science and technology, they reminded their readers, didn't always have all the answers.

Rube's Heirs

Rube Goldberg et al., spawned many later imitators. Such movies as *Back to the Future*, *The Goonies*, and *Delicatessen*, and the popular *Wallace and Gromit* series of films by British animator Nick Park, pay their own special tributes to the art of "contraptioneering." In the world of cartoons, who can forget Wile E. Coyote's doomed efforts to capture the elusive Road Runner with ever more outlandish contraptions (here the word is quite appropriate) that always backfire? A more recent application of contraptioneering was seen in a 2003 commercial for Honda cars, "The Cog," in which a Rube Goldberg machine was assembled from the parts of a new Honda model. In addition to their presence

HOW TO USE THIS BOOK

This book is divided into two parts. The first part introduces the science and principles of contraption engineering, and features general advice on components, materials, tools, and safety. The second consists of 50 contraptions for you to build or use as inspiration for your own designs. Each project is graded by level of difficulty and given an estimated build time.

All of the contraptions can be built to any scale you like—from desktop size to a basement- or garage-filler—so no exact dimensions have been listed. Suggestions for building materials have also been given, but, again, you can be flexible.

 Level of difficulty (0–10)

 Time needed

 Power

 Principles

in the media, Rube Goldberg's creations have always proved popular with amateur engineers young and old, who compete for prizes in science fairs up and down the land, and since 1988, in the National Rube Goldberg Machine Contest.

Contraption Engineering

Contraption engineering is a flight of mechanical fancy, but that doesn't mean it isn't based on some very serious scientific principles and engineering know-how.

In this section, you'll find out about the principles of classical mechanics that are the basis for contraptions, their basic components, or "simple machines," as well as guidance about tools, materials, planning, and safety.

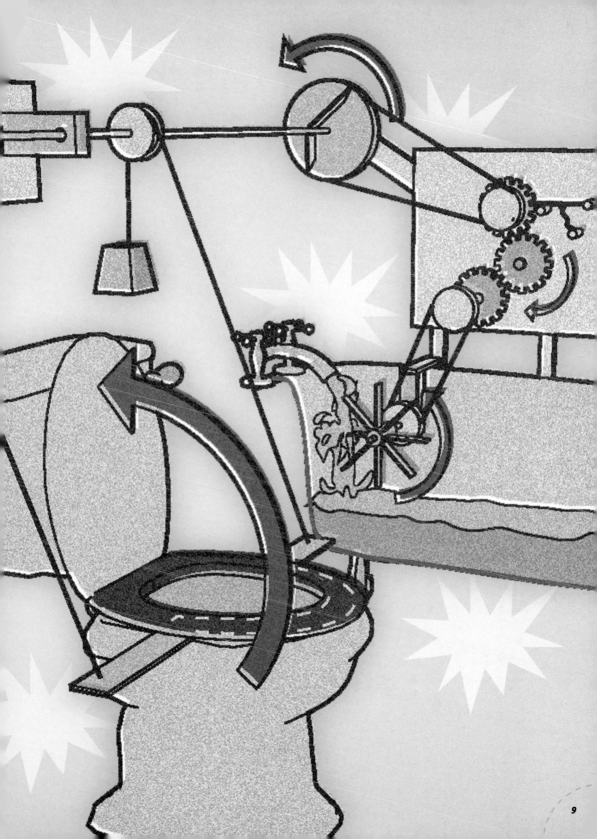

Building a Contraption

Always remember that, broadly speaking, there are only two things that go into making a great contraption....

First, it has to be able to do something. Whether it's catching a mouse, flipping a domino, or switching off a light, your contraption should be able to fulfill some sort of meaningful (if absurdly basic) task. Second, it has to do it in style. This means unexpected twists, a sense of humor, and—above all else—far greater complexity than is necessary. After all, the point of these contraptions is that they're designed to make a simple task as complicated as possible, otherwise you may just as well go out and buy a real mousetrap, apple corer, or light switch. If you can look at your design and honestly say that it fulfills both of these criteria, then you're well on your way.

To Build A Better Mousetrap

The Balls

The Trap

The Mouse

For many contraptioneers, their first introduction to the joy of contraptions was the classic game "Mousetrap®." Designed by Marvin Glass and Associates in 1963 and currently published by Hasbro. If you're unfamiliar with the original, here's how the finished trap works: turning the crank rotates a set of gears that moves the lever and pushes the "stop" sign against the shoe; the shoe upends a bucket that contains a metal ball that bounces down a staircase into and out of a pipe to hit a rod; the rod tips a bowling ball along a groove into and then out of the bath where it lands onto the diving board; the weight pitches the diver through the air into the bucket, dislodging the cage that finally traps the mouse—pure genius!

One of the joys of contraption building is being able to create something unique. You can make any number of crazy machines using the basic elements of a mousetrap.

CONTRAPTION ENGINEERING: THE RULES

1 Your contraption must work; it must work from start to finish and it must work without any human intervention. If you have to step in halfway through to give this or that component a crafty nudge, then you need to go back to the drawing board.

2 The best contraptions make each step easy to see (unless being hidden for deliberate effect); this is fine if you have a large area to work in, but otherwise you may have to compromise and ask your audience to move round the contraption as it progresses through its different steps.

3 Shake it up a bit. Instead of using a marble, use a toy car, or a cup with some toy wheels attached; hardback books make good tracks; plastic spoons have lots of uses; a drawing pin in a straw makes a great pivot . . . the list goes on.

4 Purists don't use electrical gizmos to power their contraptions–they're not banned, but they're frowned upon, especially when there are so many more inventive ways to get things going. If you're tempted by batteries, think about roller coasters–all that speed and excitement without an engine!

5 Avoid flammable or otherwise dangerous chemicals unless they're absolutely necessary. If you do plan to use anything that is potentially dangerous, be very careful to take the proper safety precautions (see pp. 24-25 for more safety tips).

6 The same applies to anything that flies. A contraption that makes a dart shoot across a room and into a dartboard is a classic, but it's also potentially lethal if you're not careful.

Getting it right

Before you start work, remember the three Ps: Preparation, Planning, and Prototype. Start by working your design out on paper, then rough it out in real life in sections, and only then start to make the finished device. It's always a good idea to start at the end, with whatever it is that you want to do–popular actions include pouring cereal into a bowl, squeezing toothpaste onto a toothbrush, unscrewing the lid of a jar, turning the page of a book, and opening a window blind.

Although it sounds back-to-front, it's important that you build from the finish backward. If, for example, a cage needs to be knocked over a "mouse," what's going to do the knocking? And what's going to cause it? Are there any other ways to spring the trap, and would any of them be more dramatic, or effective, or funny? Sometimes you suspect that some Hollywood films start life like this: with a great ending, and the hope that the rest will fall into place. You need to think the same way.

The great thing about designing contraptions is the way that you can change things around–often at the last minute–to improve the design. As long as there's enough variation in each of the individual sections, you can mix them up until you find out what makes the most exciting contraption.

Classical Mechanics

When you were in school, you might have thought that the only point to Sir Isaac Newton was to make your life miserable; alternatively, you could have thought he was the greatest genius until the man who invented the cell phone with a built-in MP3 player.

What we owe Newton, however, is nothing less than the science of classical mechanics, which neatly describes the motion of objects through space in a series of laws and mathematical equations. This is precisely the science that you'll be using when you're building and operating your own contraptions.

Sir Isaac Does Fruit

Sir Isaac Newton (1642–1727) is famous for his apple—the clue, so the legend has it, to the secrets of gravity. But Newton is far more significant than his experiences with ballistic fruit. In his *Philosophiae Naturalis Principia Mathematica* (Mathematical Principles of Natural Philosophy) of 1687 he outlined the three laws of motion that describe how all objects move (see box) and the law of universal gravitation. The "Principia" became the foundation of the science of classical mechanics, which in turn remains the basis for the branches of modern physics.

A fruity demonstration of gravity—and a possible inspiration for a contraption.

Moving Parts

Classical mechanics concerns itself with different types of motion at the macroscopic scale, that is, the world we can see, feel, and touch. When dealing with the infinitely small, and infinitely fast, you need quite different forms of physics, but when planning your contraptions, consider using different forms of motion to achieve more varied and interesting effects. Here are the main types of motion described by classical mechanics:

Translational: motion by which a body shifts from one point in space to another (for example, the motion of a projectile).

Rotational: motion by which a body changes orientation with respect to other bodies without changing position (for example, the motion of a spinning top).

Oscillatory: motion that continually repeats in time with a fixed period (for example, the motion of a pendulum).

Circular: motion by which a body executes a circular orbit about another fixed body (for example, the orbit of a planet around a sun).

A Matter of Words

In describing motion, classical mechanics employs mathematical equations, which are beyond the scope of this book to explain in detail. However, classical mechanics also uses a very precise vocabulary of terms that are often confused with more general terms. In this section, we'll look at the concepts of displacement, velocity, acceleration, and mass, which in common usage are often confused with distance, speed, and weight.

Out of Place

Displacement shouldn't be confused with distance. For example, displacement doesn't indicate how many miles a car has traveled along a winding road, but how far it has moved relative to a fixed reference point of origin along a vector (a line of direction).

Fast and Furious

In colloquial English, speed is often used to mean velocity and acceleration, which are the two key concepts of motion. Velocity is the change in an object's position (its displacement) over time. Velocity can be either positive or negative, depending on the direction of motion. The conventional definition of speed is that it's the magnitude of velocity, and an object can't have a negative speed. Acceleration is the rate of change in the velocity over time. It can arise from a change with time of the magnitude of the velocity, or of the direction of the velocity, or both.

Putting things in their place

A few simple formulae to get you going

Displacement
= straight-line distance from origin

Speed
$$= \frac{distance\ traveled}{time\ taken}$$

Velocity
$$= \frac{displacement}{time}$$

Acceleration
$$= \frac{change\ in\ velocity}{time}$$

POETRY IN MOTION: NEWTON'S THREE LAWS

1 *A body continues in its state of rest or of motion in a straight line unless it's made to change that state by a force applied to it. In other words, an object can be in one of two states: at rest or in motion. On earth, a moving object is stopped by friction, usually from the ground, water, or air.*

2 *The change of motion of an object is proportional to the force applied to it, and is made in a straight line in the direction in which the force is applied. In other words, the force exerted on one object by another is equal to the object's mass times its acceleration.*

3 *To every action there is always an equal and opposite reaction; that is, the reciprocal actions of two objects on one other are always equal and directed to opposite parts.*

Weighty Matters

The terms mass and weight are often confused. However, in classical mechanics their meanings are quite different. An object's mass is a measure of its inertia; that is, its resistance to deviating from uniform straight-line motion under the influence of an external force. Weight is simply the force exacted by the gravity of the earth with which the earth attracts an object.

The Building Blocks: Basic Components

Although it's a good idea to start from the end and work backward, there is no set way of designing a contraption. You could find an intriguing component and build a device around it, get a really wacky idea for a task to accomplish, or just play around with different components and start putting them together into a pleasing array.

Whichever way you dream up your contraption, sooner or later you'll have to sit down to design and draw it, scale its components so they all fit together, and write out a list of materials and the tools you'll need for the job.

On the Drawing Board

You may have a great idea: perhaps a contraption on which you place a bottle that tilts to fill a glass, which it then pushes within reach of your amazed dinner guest. Brilliant! How many steps is that? Three? Look again. The bottle must fill the glass, but not overfill it, it then has to tilt back into position; the mechanism pushing the glass forward has to retract for the next serving. And you could make it a lot more complicated: moving the bottle, glass or liquid in other interesting ways; uncorking and re-corking the bottle; dropping an olive or ice cube into the glass. You'll probably get a million ideas all at once—too many to remember, so start by jotting them down and sketching them out on a drawing pad. But a sketch won't be enough. The next step is to draw it out.

Squares, Scales, and Splines

Technical drawing is a science—even more so now with computer-aided design programs—but you're not drawing design specs for a production line, just something that will make sense to you. It can be rough, it can be ready, but it does have to be to scale, so when it comes to sourcing or making components, you know what you aiming for. The best way to scale a drawing is to use graph paper, on which you can decide whether a square stands for 1 in, 1 cm, or 1 mm. Then all you need is a ruler and a decent (thin-nibbed) drafting pen that'll make neat lines and won't smudge over several squares.

Also useful: a compass, not just to draw circles, but to measure between two points where you can't easily use a ruler, French curves and a spline (a bendable ruler) to draw curves and irregular shapes, a protractor and set square to measure and draw angles, and if you're feeling really technical, an engineer's or architect's scale ruler. And whatever unit of measure you use, remember what happened to the Mars Climate Orbiter when the engineers mixed up imperial and metric measurements—Ooops! $125 million down the drain.

Material Needs

Each contraption is designed so that the materials used to make it are interchangeable. We've made some helpful suggestions, but in a lot of cases you can use whatever's available. However, there are some general materials that often prove useful, such as: assorted dowels; nails; screws; fixings; nuts and bolts; wire and string; hinges; joining plates and brackets; and various lengths and thicknesses of timber, metal rods, plastic sheeting, and thick cardboard.

Although you can go to hobby stores and specialty suppliers on the Internet for components, one of the joys of contraption building is improvising with what you have at hand and recycling discarded components from other devices. Here are a few ideas for useful odds and ends to be on the lookout for use in the components that make up the contraptions: marbles; ball bearings; golf balls; ping-pong balls and balloons; springs and wire from spiral-bound notebooks; plastic containers; cups; plastic bottles; modeling putty; cheese (yes, cheese!); rubber bands; pulleys, magnets, and weights; toy cars; wheels; bells; fizzy tablets; bathtub toys . . . Get the idea?

GETTING TOOLED UP

No two contraptions require exactly the same tool set, not least because the construction materials vary. Try to put some thought into the tools that you're going to need. Screwdrivers, knives, scissors, pliers, wrenches, hammers, an electric drill, adhesives, lubricant, and a wood saw are all fairly standard items, but here are a few less common tools that may be useful:

- *A carpenter's square, tenon saw, and miter box will enable you to make accurate straight and angled saw cuts.*
- *Specialty saws and cutters for glass, plastic, and metal.*
- *A hand drill with wood, metal, and glass bits for close, smaller work where an electric drill would be too powerful or cumbersome (and think of your "carbon footprint").*
- *An awl.*
- *A selection of chisels for cutting joints and rebates.*
- *A selection of vices and clamps for assembly and gluing.*
- *A plastic hammer or wooden mallet to tap in without marking a fragile surface.*
- *A bubble level (very important).*
- *A spokeshave, for shaping curves in wood.*

Simple Machines

The word "machine" suggests an engine, and many interlocking components; however, a "simple machine" has no independent power source and few or no moving parts. Described here are six of the most common and most useful simple machines.

The trick is to combine a few simple machines to produce more interesting effects. Although other components are involved, simple machines provide the core of a successful contraption.

Levers

There are three types, or "classes," of levers. In each there is a fulcrum (F), a load or resistance (R), and an effort (E). The class of lever varies with the relative positions of the FRE. In a first-class lever, the fulcrum is between the resistance and the effort. The clearest example of a first-class lever is a seesaw. In a second-class lever the resistance is in-between, as in a door. Third-class levers have the effort in-between, as in a spring-loaded mousetrap. Levers are useful to change the focus and direction of your contraption; experiment by using different classes of levers, weights, and forces to achieve unusual effects.

Inclined Planes

In simple terms, these are sloping surfaces like ramps. Use them to help rolling objects pick up speed, so they can apply more "kinetic energy" when they hit something else. Alternatively inclined planes can be used to lift something from one level up to the next instead of pulling it straight up. It's easier, for example, to roll a heavy object up a ramp and into the back of a truck than it is to lift it straight up with a crane.

Pulleys

These are wheels with a groove around the edge designed to take a rope or string. When you attach an object to one end of the rope and pull on the other, you'll find it's easier to lift the weight than it is if you tried to pull it straight up from above, and the more pulleys you use to lift the weight the easier it gets. This is because of the improved mechanical advantage (see box).

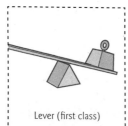

Lever (first class)

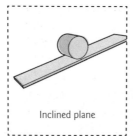

Inclined plane

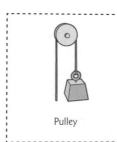

Pulley

Screw

Screws

A screw is basically a sloping plane that's wrapped around a cylinder. This allows the screw to move itself backward and forward through something more easily, or allows other things to rotate around the thread. By using a variation like a spring, you can achieve all sorts of interesting results. The screw is one of several inventions (another one was the lever) credited to the Greek inventor Archimedes (287–212 BCE). However, archeological evidence suggests ancient Mesopotamians were already using screws some 300 years before his birth.

Wedges

These are two inclined planes joined together to form a shape like a piece cut from a wheel of cheese—which is ironic really, when you consider that some of the most famous contraptions are designed to catch mice. But rather than being used as bait, wedges are normally used to separate objects by being driven into them, or into the gap between them. Curiously, the most common use for a wedge in the real world is to keep a door open, which actually makes use of the friction caused as it tries to push the door and the floor apart in order to the keep the door in place, rather than actually separating them. The axe is a more classical example of a wedge.

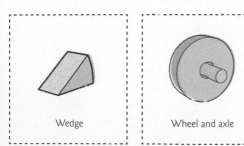

Wedge · Wheel and axle

GETTING THE ADVANTAGE

Mechanical advantage (MA) is the factor by which a machine multiplies the force applied to it. In other words, how much more easily and faster a machine will run. To work out the MA of a machine, you divide the resistance force by the effort force. For example, to work out the MA of a first-class lever, divide the effort arm length by the resistance arm length. For an inclined plane, the MA is the plane's length divided by its height. Two simple machines that are particularly good at providing MA are the wheel and axle, and the pulley.

$$MA = \frac{load}{effort}$$

Wheel and Axle

The wheel rotates around a central point, or axle. Because the circumference of the edge is greater than that of the axle, a short powerful force at the axle will move the wheel a much greater distance, making it an efficient way of moving objects (mechanical advantage again). The mechanical advantage is equal to the circumference or radius of the wheel divided by the circumference of the axle. Mother Nature never invented the wheel, making it a uniquely human creation—although the Aztecs, Inca, and Maya, despite their advanced mathematics, never created wheeled vehicles.

Wheels, Cogs, Gears, and Pulleys

In this and the next section, we'll take a closer look at the individual components that are used in the contraptions in this book. Focusing on circular and rotational motion, this section covers wheels, cogs and gears, and pulleys and examines how they illustrate the concepts of simple machines and mechanical advantage.

Round and Round

The components we're looking at in this section all apply the same wheel-and-axle principle described in "Simple Machines" (see pp. 16–17). The wheel, used in forward motion or to perform rotational work as in a windmill, makes full use of its mechanical advantage. In Contraption NO.36, the "Water-Powered Bath Alarm" (see pp. 100–101), you'll see a couple of applications of the wheel: cogs and gears and two rotating wheels, one turned by water just like the paddlewheel of a watermill.

Putting Your Teeth on Edge

In the simplest terms, a gear consists of two wheels with teeth along their edges turning one another—either directly because the teeth fit together or via a chain, as in a bicycle. The bicycle is one of the best examples of gearing that is visible in everyday life. When you pair

gears of different sizes you gain mechanical advantage, in which the larger gear turning slowly makes the smaller gear turn faster. However, this isn't magic: no matter how the gears are arranged, you can never get more power out of them than you put in. But by combining large and small gears correctly, you can produce more torque (twisting force).

Gears are also sometimes called "toothed wheels," "cogged wheels," or just simply "cogs;" the teeth on the gear are also sometimes called "cogs." The smaller gear in a pair is known as the "pinion," and the larger, either the "gear," or the "wheel." The interlocking of the teeth in a pair of meshing gears means that their circumferences must move at the same rate of linear motion. The rotational speed is proportional to a wheel's circumferential speed divided by its radius, so the larger the radius of a gear, the slower its rotational speed, when meshed with a gear of given size and speed. We've used cogs and gears in several projects. For example, Contraption NO.21, the "Magneto Sugar-Cube Dispenser" (see pp. 70–71), uses cogs to translate the motion from the weight dropping to the horizontal movement of the crane; similarly, in Contraption NO.36, the "Water-Powered Bath Alarm," the cogs moving the arms carrying the bell reverse the direction of the movement.

The mechanical advantage of a gear is its "torque ratio." The torque ratio can be worked out by looking at the force that the tooth of one gear exerts on the tooth of the other. Consider two teeth in contact at a point on the line joining the shaft axes of the two gears. The torque is equal to the circumferential component of the force multiplied by the radius. Thus we see that the larger gear experiences greater torque and the smaller gear, less. The torque ratio, therefore, is equal to the ratio of the radii of the gears.

Pulling Your Weight

A pulley is a wheel with a grooved edge, or "sheave," which holds a rope or string. Pulleys are used in groups to reduce the amount of force needed to lift a load. However, the same amount of work is necessary for the load to reach the same height as it would without the pulleys—it just feels a lot easier. The magnitude of the force is reduced, but it must act through a longer distance. The effort needed to pull a load up is roughly the weight of the load divided by the number of pulleys. The more pulleys there are, however, the less efficient the system becomes, because

OTHER COMPONENTS

Although it's possible to buy components for the contraptions shown here from specialty toy and hobby stores—or on the Internet— these aren't the only options open to keen contraptioneers. For a start, many off-the-shelf wooden toys and building sets can be pressed into service; or, if you have the time and skill, you can make your own components. Another good source of materials are to be found in your home, in discarded white and electronics goods. And don't neglect junkyards, second-hand stores, and yard sales as good sources of ready-made components. Keep your eyes open—an unusual component can become the centerpiece of a really interesting contraption.

there is more friction between the rope and the pulleys. No one knows who invented the pulley, but the first documented block-and-tackle system for lifting heavy loads was designed by our old friend Archimedes. The most pulleys are used in Contraptions NOS.46–50 (see pp. 116–125), which use them to achieve a variety of effects. As you can see, pulleys are used here to lift loads, and also to translate motion from one plane to another—in other words, from vertical to horizontal.

Pendulums, Projectiles, Pivots, and Springs

This second roundup of contraption components features a selection of simple machines—the pendulum, pivot, projectile, and spring—that use a variety of classical mechanical principles to store energy, move objects, or exert force over a distance.

King of the Swingers

The gravity pendulum or "bob" pendulum consists of a weight, or bob, on the end of a string or wire, which, when given an initial push, will oscillate back and forth under the influence of gravity over its lowest point. The French physicist and inventor Léon Foucault (1819–1868) built the best-known pendulum as an experiment to demonstrate the rotation of the earth. His 1851 pendulum used a 61 lb 11 oz (28 kg) bob suspended at the end of a 220 ft (67 m) wire whose direction changed as the earth rotated. Contraption NO.21, the "Magneto Sugar-Cube Dispenser" (see pp. 70–71), uses an application of the pendulum that was popular as an executive toy in the 1970s known as "Newton's Cradle." The cradle demonstrates two fundamental physical principles: the conservation of energy and the conservation of linear momentum. In the former, energy in an isolated system can't be either created out of nothing, or destroyed; it can, however, be converted from one form of energy into another; and the latter, that the total momentum of such a system is constant; hence, without the action of air friction, the balls in the cradle will continue moving at the same velocity.

Fire!

A projectile is any object sent through space by the application of force. There are three common types of projectiles: dropped objects, objects thrown vertically upward, and objects thrown upward at an angle. Once launched, a projectile continues in

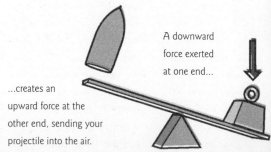

A downward force exerted at one end...

...creates an upward force at the other end, sending your projectile into the air.

motion by its own inertia and is influenced by the downward force of gravity. If there were any other forces acting upon it (in other words, if it had its own power source, like some kind of engine), then that object wouldn't qualify as a projectile. The most common application of projectile physics is in "kinetic energy weapons," such as bullets fired from a gun. Contraption NO.31, the "Wind-powered Rocket Launcher" (see pp. 90–91), uses an elastic band as the power source to launch a projectile upward into a waiting container.

Balance of Power

Simple pivot devices such as seesaws and old-style weight scales use the principle of leverage. These devices are first-class levers (see pp. 16–17) in which the fulcrum, or pivot, is located between the effort and load (or resistance). In classical mechanics (see pp. 12–13), a lever is any device in which a rigid bar with a pivot point multiplies the mechanical force that can be applied to another object. The principle of leverage is derived using Newton's laws of motion, which specify that the amount of work a lever can do is equivalent to the force multiplied by the distance. For example, to use a lever to lift a unit of weight with an effort of half that unit, the distance from the pivot point to the spot where the effort is applied must be twice the distance between the weight and the pivot point. So to halve the effort of lifting a weight resting 4 in (10 cm) from the pivot point, we need to apply force at a point 8 in (20 cm) from the other end of the pivot point. In practice, the force is applied by pushing down at one end, which causes the device to swing about the pivot point, overcoming the resistance at the opposite

end. Contraption NO.6, the "Coin-Operated Toothpaste Squeezer" (see pp. 40–41) uses a double-pivoted bar to get the ball (or here, the coin) rolling. Contraption NO.16, the "Tabletop Fountain" (see pp. 60–61), uses a slightly more complicated application of the pivot in the shape of weight scales.

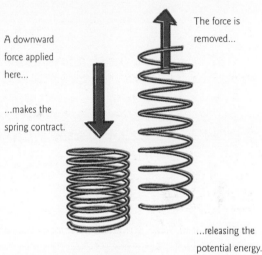

A downward force applied here...

...makes the spring contract.

The force is removed...

...releasing the potential energy.

Spring

The spring we're most familiar with is the "coil" or "helical" spring that is made by winding a wire around a cylinder. This is a type of "torsion spring" because the wire is twisted when the spring is compressed or stretched. Less familiar are compression springs that are designed to become shorter when loaded. Their turns do not touch when they're in an unloaded position. A "volute" spring is a compression spring in the form of a cone. When compacted, the coils aren't forced against one another, thus permitting greater movement. In Contraption NO.41, "Energy-Saver Light Switch" (see pp.110–111), such a spring is used to store energy that is released to turn a pulley mechanism.

Magnets and Other Forms of Propulsion

Although it's permissible to use a battery- or mains-powered electric motor in contraption engineering, it somehow misses the spirit of the art; similarly, just giving the first component a hefty whack to get the device going lacks elegance. Fortunately, there is no shortage of methods for powering your contraptions. Here are a few examples to get you started.

Air Superiority

There is no need to mess around with oxygen cylinders to use air pressure; all you require is a bicycle pump or a balloon. Used as a component in your contraptions, a pump will compress the air, giving it quite a kick, while a balloon will store air, to be released when and where you need it (you can also burst a balloon if you want a big finish). Air pressure in the form of wind is an easily obtainable resource. Create an air current by opening two windows to send a model sailboat across a bowl of water, or to power a small windmill (and if the wind fails, we'll allow you an electric fan).

Heavyweight

Where would we be without gravity? On our way to the Crab Nebula and accelerating, most likely! But in contraptioneering gravity has numerous applications. Once you've built in height,

gravity will do the work for you; the trick is to use it creatively. A ball or cylinder running down a track, a weight on the end of a pulley, ball bearings, water, or sand filling a container, or a weight sliding down an incline—the possibilities are too numerous to list them all.

Fluid Mechanics

Liquid is an extremely versatile power resource. You can float things on it, you can use it as a weight in a gravity device, you can change its pressure, and you can make it run through pipes and tubes to get it to where you want it to be creatively, and with that all important element of surprise. Water is an obvious choice, but don't overlook the properties of other liquids: the viscosity of oil, the volatility of white spirit, and the flammability of alcohol (in small amounts and with great care), can all be employed to good effect.

Power of Attraction

Magnets provide a secure but temporary coupling between two metal surfaces. They can be hidden by cardboard, paper, or plastic, and still exert their power while being invisible to the observer. Naturally occurring magnetized minerals are known as "loadstones" and were used to power the earliest compasses; however, magnets are now manufactured. You can magnetize a piece of metal by running a magnet over it a few times (in a consistent direction).

Playing "Hookey"

A spring is a flexible, elastic object, often made of steel, which is used to store mechanical energy. The properties of springs were described by the seventeenth-century physicist and polymath Robert Hooke (1635–1703), and are summed up in the law that bears his name. According to Hooke's law, the force that the spring exerts on an object is directly proportional to its extension, and always acts to reduce this extension. Springs of different types are found in many simple mechanisms with moving parts.

Twister

When you twist a rubber band fixed at both ends, what you're really doing is stretching it. This builds up energy, which it wants to release by untwisting itself so that it can return to its natural tension. The best-known application for this very simple technology is probably the rubber-band-powered model airplane. It was nineteenth-century French aviation pioneer, Alphonse Pénaud (1850–1880), who made the first rubber-band-powered model plane, which he christened the "Planophore." You can follow Pénaud's example and use a rubber

WIND-UP

Simple clockwork mechanisms can be found in many old toys, like wind-up robots and cars. Typically, clockwork uses a winding device to store power in a spring. The power is released gradually through the use of wheels, whether linked by friction or cogs, which can direct motion or gain speed or "torque" (see pp. 18–19).

band to power the propeller of an airplane, or a windmill, or perhaps even the wheels of a toy car.

Some Like it Hot

Temperature is a ready source of motive power for your contraption. In the form of heat, it can displace the air under the blades of an "Angel Chime" to make it spin. Heated water creates steam, which can be harnessed to power your contraption. However, any device with a naked flame or boiling water needs to be treated with caution. Keep a fire extinguisher handy. Remember that temperature can mean both hot and cold. A "Drinking Bird" is another application of heat–cold exchange; in fact, it's a basic heat engine that takes advantage of the pressure differential in two glass bulbs filled with dichloromethane gas to make a toy bird dunk into a glass of water. Ice can also give you some interesting effects, as the melted water can be redirected into a bucket in a gravity device. Melting ice also provides a simple and automatic time delay to start your contraption.

Safety and Other Considerations

When thinking about safety, you're not just thinking about your own life and limb, but also the safety of anyone else in the vicinity.

Remember that not everyone is as safety conscious as they should be: a small child might want to handle all those brightly colored tools and taste some of those interesting-looking chemicals, an elderly person might be wobbly on their feet, and, last but not least, a pet can get into all sorts of trouble. The advice given here should be common sense, but you'd be amazed at how many people are injured every year for not doing the obvious and using common sense.

Contraption Sense

When you design your own contraption, make sure that the finished product and its component parts are safe—keeping in mind inquisitive small children and pets. Projectiles and pendulums can be fun, but make sure they can't go astray and hit an unwary passerby. Ensure that the materials you have employed are "fit for purpose," that is, that they're strong enough for the forces that they're exposed to and expected to withstand to work your contraption. If in doubt, run a series of tests.

Clothes Sense

When working on your contraption, wear protective gear: overalls to protect your clothes and any exposed area of skin from cuts, abrasions, and chemicals; goggles when sawing, drilling, and sanding; a dust mask when handling chemicals, sawing, sanding, and spraying; boots to protect your feet, and rubber-soled shoes when working with power tools. Avoid wearing loose clothing or jewelry that might get caught up in the moving parts of drills, routers, jigsaws, and other tools.

Tool Sense

Plan ahead and make sure that you have all the tools you need for the task at hand. It'll be faster, safer, and above all, you'll get better results. Store your tools in a toolbox or rack out of reach of kids and pets. Get into the habit of putting away your tools immediately you've done using them. That way, you'll always know where to find them, and you won't injure yourself by accidentally brushing against them or stepping on them. Use the cover guards provided with sharp tools such as saws and chisels. When using knives, planes, chisels, and saws, always work away from you. If you need to work on the other side of a piece, turn the piece, not the tool. Treat power tools with respect: turn the power supply off at the socket to change drill bits and saw blades, and engage any safety guards and catches provided. Finally, keep all tools, manual or power, clean and in good repair. If they need to be fixed, take them to a professional or replace them with new ones.

Work-Area Sense

If your home isn't blessed with a garage, basement, outbuilding, or spare room that you can convert into a permanent contraption-making workshop, it's even more important to have a well-organized workspace. The contraptions in this book aren't large, so you won't be sawing whole tree trunks, but you'll need a clear area to work in, 360-degree access to the piece you're working on, and a level, stable, and solid surface to place it on. A good piece of kit to invest in, if you haven't got one already, is a foldaway workbench with a built-in vice. You'll also need a clear table-top where you can lay out any pieces to be assembled, glued and clamped, or painted. Another important work-area requirement is storage for your tools and materials. Tools need to be kept out of harm's way (both yours and other people's) and easy to find when you need them. Materials need to be kept clean, dry, and safe from damage. Finally, your work area should be well ventilated, to vent fumes from paints, glues, and solvents, and well lit to make it easier for you to work.

If you decide to build a large-scale project, watch out for overhead objects.

Shopping sense

The joy of contraptions is in the making. If all you wanted was a mousetrap, you could go out and buy one cheaply in any hardware store. But building a mousetrap that involves half-a-dozen pulleys, a couple of levers, and a falling ball to make it work, is a joy forever (or until you take it apart to build the improved model). Depending on your level of skill, you could manufacture all the components of a contraption, metal-work and all. However, you have to be realistic about your abilities when you start out. Drilling and sawing wood is one thing; metal and glass are quite another. So make all the parts that you can make safely, and buy or find the rest. Before setting off to the store or looking on the Internet for specialist parts, check out what you have at home. It's amazing what you can find: discarded parts from bicycles to outboard motors, and from radio innards to old kitchen equipment. Instead of buying new, reuse and recycle by going to junk yards, garage sales, and second-hand stores.

Environmental Sense

We all have to think about the environment these days, and contraption building as a pastime has strong eco-friendly credentials. If you use hand tools, recycled materials, and don't power your contraptions with electricity or batteries, they will have a minimal carbon footprint. So you can have fun, entertain your friends, and save the planet at the same time.

CONTRAPTIONS 1-5

If like me, you or your partner likes to multitask when you're reading, then the contraptions in this section could be the answer to your prayers.

Got to that tricky part in the instruction book or manual and need to turn the page, but got both hands full?

What could be simpler than to use a contraption-engineering solution to turn the page? Almost anything!

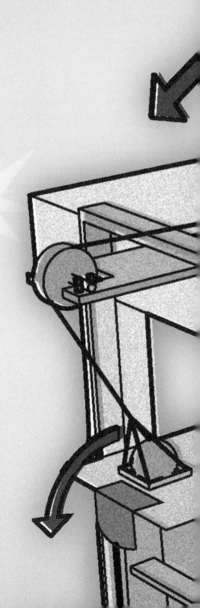

Hands-Free Page Turner Projects

The golden rule of contraption engineering is: If a task is worth doing, it's worth doing in the most complex way imaginable. There are few tasks less complicated than turning the page of a book, but with hands-free page turning, reading has never been so challenging....

 Power: *air pressure, gravity*

 Principles: *pulley, projectile*

> **Suggested Components**
> wood • hinge • thick cardboard • plastic bottle • wine-bottle cork • balloon • rubber tubing • cotton spool • binder clip • golf ball • string • modeling putty • ballpoint pen • and a book…

The Inspiration

Appropriately for a book on contraption engineering, the inspiration for the first section was the book itself. And as well as making the simple, mind-blowingly complex, the "Hands-Free Page Turner" does have a point. It's something that's happened to each and every one of us. You're hard at work reading a book—a reference book, or an instruction manual—and then comes that moment when you've got a screwdriver and a hammer in one hand, a length of timber in the other, a pencil between your teeth, and you need to turn the page to see what to do next. You look around and the only other "person" in the den is the dog who looks up at you balefully, asking if it's not time you stopped messing around and took him out for a walk. No help there. That's

where the "Hands-Free Page Turner" comes into its own—as long, of course, as you only want to turn one page at a time; but, hey, if we handed you everything on a plate, where would be the challenge?

Rules and Levels

We've chosen to start with the "Hands-Free Page Turner" "because the basic model is a relatively simple design that only has four main components: the "Balloon Air Pump," the "Bottle Cannon," the "Gravity Pulley," and the page-turning "Roller Arch" itself. Of course, as you'll see from Contraptions NOS.2–5 (see pp. 32–35), you can add and take away components and complexity, but we've rated this design as a 3 out of 10, which you should be able to put together and get operational within half a day. As such, it's a good project for the novice to start getting to grips with the principles of the artful science of contraptioneering. But the cool thing about this project is that it illustrates another rule of contraption engineering (the silver one maybe): Make it easy to see what's going on.

Bits and Pieces

The "Hands-Free Page Turner" is made entirely of items you'd be able to find in the average home and garage. It doesn't use any complicated gadgets, gears, or gizmos that you'll need to buy from a hobby shop or order online and wait three weeks for. So another two rules of contraption engineering that Contraption NO.1 illustrates are: Improvise and Recycle (or "make and mend" in old-fashiioned terms). The air pump's foot pedal is made from a few off-cuts of timber hinged together (you might want to add some felt between the wood and the balloon in case of splinters); the cannon is a plastic water or juice bottle with its end cut off and a

The "HandsPage Turner" uses a "Roller Arch" to help guide the string and turn the page in a smooth arc.

bottle cork; the "Gravity Pulley" could be a cotton spool mounted on a bolt axle, and the weight a golf ball; and the arch is any flexible material—we've used cardboard, but plastic or wood would do. Add a lot of string and a binder clip and you're ready to rock, or, in this case, turn a page.

Power and Principles

The initial power for the "Hands-Free Page Turner" is provided by a simple application of air pressure—namely, compressing the air in a balloon. The compressed air forced through the narrow tube will increase its force, and the cork blocking the other end becomes a pressure valve. You should get quite a kick when the cork becomes a projectile (see pp. 20–21). The cork triggers the second form of stored energy in the device: the mass of the golf ball, which is about to be exposed to the downward tug of gravity. The single-pulley system changes the direction of the force from vertical to lateral, and it improves the mechanical advantage (see pp. 16–17) of the device. Naturally you'll have to experiment with the length of the pulley arm to get a smooth pull. You don't want your page to stop halfway, or worse, be ripped out of the book.

Variations on a Theme

Contraption NO.2 (see p. 32) gives you a completely new outcome, though like the other projects in this part it retains the "Roller Arch" mechanism. Contraption NO.3 (see p. 33), NO.4 (see p. 34), and NO.5 (see p. 35) return to the page-turning theme, but change the motive power involved.

CONTRAPTION NO. 1

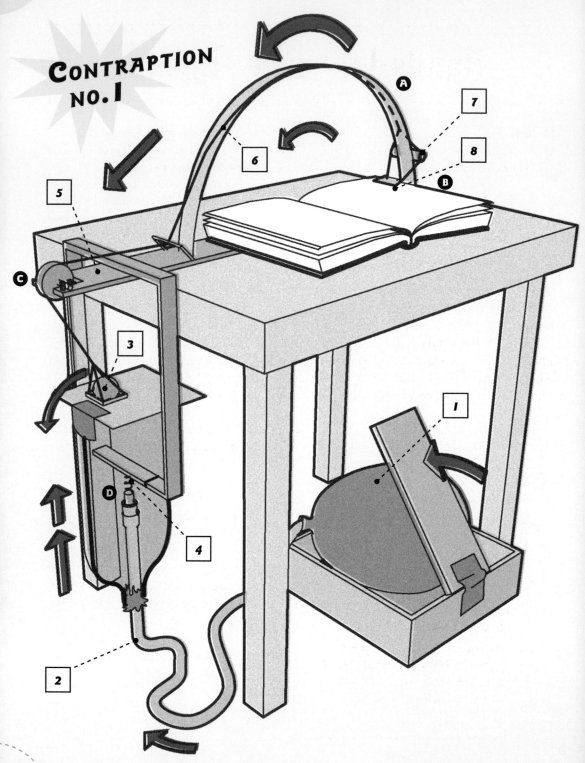

Hands-Free Page Turner

 Level: 3

 Time: *half a day*

You'd be able to find all of the components for this contraption in the average home and garage, and be able to buy anything that's missing from a local store. In terms of skills, you'll be cutting and assembling rather than performing tricky engineering, as few of the components need elaboration. The same is true for the tool kit you'll need: a knife to cut the cardboard, a wood saw, a screwdriver, a hammer, and adhesive. As such, it would be a good project on which to start a young contraptioneer.

1 The device kicks off with air compression. Don't overfill the balloon otherwise it'll probably burst. You could also pad it with some felt to guard against splinters. The air pressure increases as it's forced through the tubing.

2 The air pressure builds up in the tube until it's strong enough to dislodge the cork closing the other end of tube, forcing the pressure valve to open.

3 We've used a golf ball, but you might want to use a rubber ball or a small rock.

4 The cork hits the cardboard flap on top of the bottle, and knocks off the golf ball, which is suspended in a rope cradle. The downward force of the weight is translated into a lateral pulling force on the string by a pulley.

5 The pulley arm is firmly attached to the table, and is made of a rigid material such as wood, so as not to bend under the load.

6 The "Roller Arch" needs to be made of a flexible but fairly rigid material. It could be wood, thick cardboard, or plastic. We'd recommend plastic as it produces less friction.

7 The roller itself, which is made of a ballpoint pen sawn in half, is attached to a clip that holds the page to be turned. As the ball descends, the roller is drawn along the arch, turning the page smoothly.

8 Use a clip such as a paperclip, binder clip, or bulldog clip to attach the page to the roller.

CONTRAPTION CONNECTIONS

The "Roller Arch" **A** can be made from a strip of smooth plastic. The arc needs to trace the trajectory of the edge of the turning page **B** too high, and you'll rip the page out of the book! Place the top pulley **C** directly in-line with the path of the arch, so that the falling golf ball pulls the roller across the apex of the arch. The cork **D** should be tight enough to allow the pressure to build up as you squeeze the balloon, but not so tight that the balloon explodes before the cork fires. Make sure that all the tube connections are airtight, and use flexible plastic or rubber tubing.

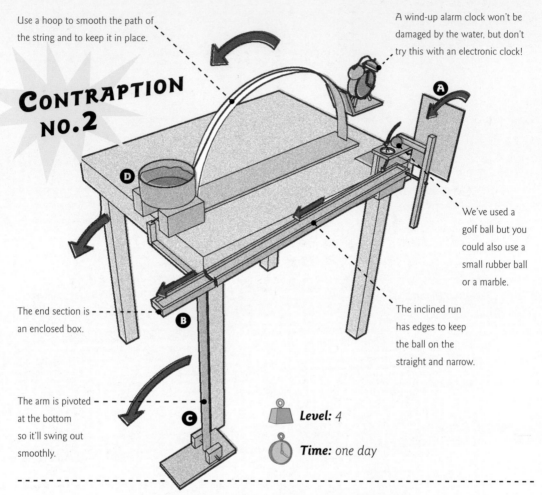

Use a hoop to smooth the path of the string and to keep it in place.

A wind-up alarm clock won't be damaged by the water, but don't try this with an electronic clock!

A

D

We've used a golf ball but you could also use a small rubber ball or a marble.

CONTRAPTION NO.2

The end section is an enclosed box.

B

The inclined run has edges to keep the ball on the straight and narrow.

The arm is pivoted at the bottom so it'll swing out smoothly.

C

Level: 4

Time: one day

Alarm Clock Dunker

Proving that inspiration can come from anywhere, we got the idea for how to power this contraption from the start run of a pinball machine. When you sleepily push the wooden flap **A** with your hand, it knocks the ball (use a fairly heavy one) down through a hole onto an inclined plane (see pp. 16–17); it picks up speed as it rolls down, slamming into a dead end section **B**

that is attached to the end of a pivoting arm **C**. The extra weight knocks the arm away from the table, drawing the string attached to the roller to pull the alarm clock up and over the arch and into the water container **D**, which will hopefully silence it.

Don't forget to place the contraption within easy reach of your bed.

CONTRAPTION NO.3

High Roller

We had a lot of fun working our way through this one. We've reused the air-pressure application from Contraption NO.1 (see pp. 30–31), but in a different way. When you blow into the tube **A**, the air pressure sends a needle-tipped cork projectile **B** into the balloon. Bursting the balloon frees the large wheel and axle (made of cardboard or plastic) and sends it down an inclined plane, pulling the roller with it. Both the inclined plane **C** and the base of the triangle **D** have a slit through the middle in which the wheel rolls. The base should be wood, but the inclined plane and upright piece can be cut from cardboard.

 Level: 3

 Time: less than half a day

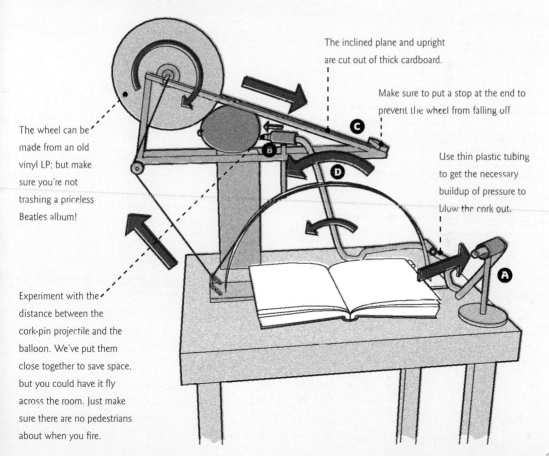

The inclined plane and upright are cut out of thick cardboard.

Make sure to put a stop at the end to prevent the wheel from falling off.

The wheel can be made from an old vinyl LP; but make sure you're not trashing a priceless Beatles album!

Use thin plastic tubing to get the necessary buildup of pressure to blow the cork out.

Experiment with the distance between the cork-pin projectile and the balloon. We've put them close together to save space, but you could have it fly across the room. Just make sure there are no pedestrians about when you fire.

33

Level: *2*

Time: *2 hours*

CONTRAPTION NO.4

Can-tilever

There are only three components to this simple contraption: a "Coffee-Cup Crane," a "Screw Brake," and the "Roller Arch." The device starts when you put a can of soda or a coffee-cup on the suspended shelf **A**, which can be made of wood or particle board. The weight of the object pulls the string down, the force being translated by the pulley **B** at the top of the crane arm. The string passes through the "Screw Brake" **C**, which is a wooden block that presses on the string and slows its movement with friction. You can tighten or loosen the screw to adjust this to the right tension.

Use a couple of thin layers of rubber between the blocks of the "Screw Brake." The brake should be set to take account of the weight of the unloaded shelf, so that the device is held in balance until you add the cup.

Instead of a screw, you can use a bolt and a butterfly nut. This will make it easier to adjust the device for different weights.

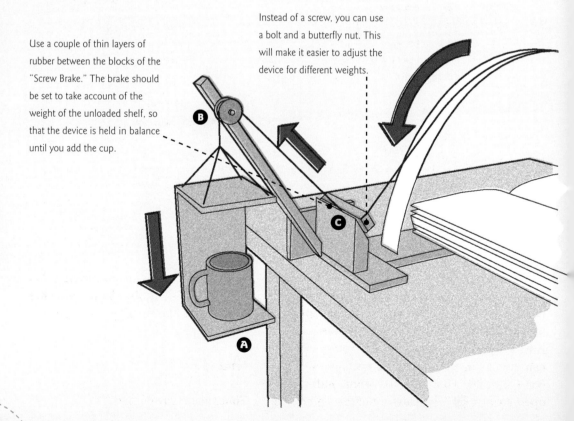

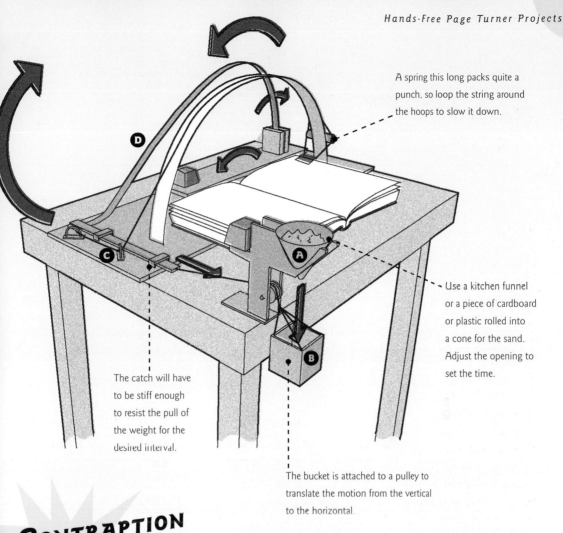

A spring this long packs quite a punch, so loop the string around the hoops to slow it down.

Use a kitchen funnel or a piece of cardboard or plastic rolled into a cone for the sand. Adjust the opening to set the time.

The catch will have to be stiff enough to resist the pull of the weight for the desired interval.

The bucket is attached to a pulley to translate the motion from the vertical to the horizontal.

CONTRAPTION NO.5

Sands of Time

This project gives the reader a set amount of time to finish the page. When the plug (a cork will suffice) at the bottom of the funnel **A** is removed, the sand starts to run down into the container **B** below, you can use a can for this. Once enough sand has fallen, the container descends, pulling open a catch **C**. The catch releases a giant spring **D** made from a strip of metal, this flips up behind the "Roller Arch," drawing the string with it and turning the page.

Level: *4*

Time: *half a day or more*

CONTRAPTIONS 6-10

The bathroom is usually the most functional room in the house. And unless you're a big fan of potted plants, candles, and other "chichi" accessories, there isn't much in the shops designed to make it more stimulating. But as you'll be spending at least an hour in there every day of your life—showering, shaving, and the rest—why not inject a little bit of fun into this antiseptic environment?

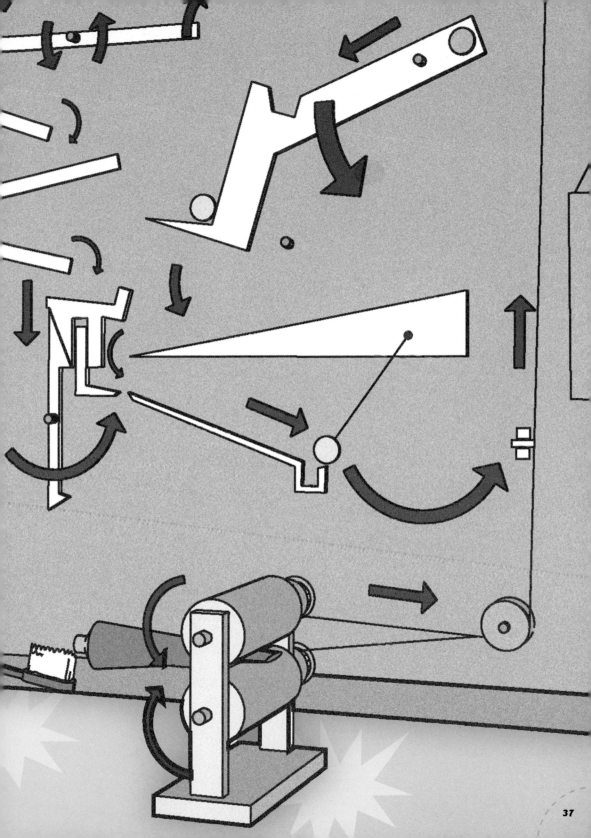

Coin-Operated Toothpaste Squeezer Projects

As rooms go, the average bathroom is functional, bare, boring, and cold. Yet if you do the math, you'll find that, over a lifetime, you're going to be spending an average of 1,200 days bathing, shaving, and so on. So why not put some fun into what is often a dull, repetitive routine? Enter the "Coin-Operated Toothpaste Dispenser," a contraption that can make cleaning your teeth fun for adults and kids alike, as well as providing an original ablutionary art installation that will brighten up your bathroom.

 Power: *gravity*

 Principles: *lever, pulley, pendulum*

The Inspiration

Not all of us are good in the mornings; and the glare of the fluorescent tube that illuminates the reflection in the mirror isn't always a pretty sight. And not every day is a gloriously sunny summer's day, that inspires us to go out to face the day's challenges with a song in our hearts. In short, we all have days when we need a little help to make sense of it all, to give us some perspective, as well as an antidote to boredom and early morning anxiety. And yet, what is there in the average bathroom to give us that little lift: the drab porcelain and chrome fitments, the antiseptic tiles,

the cold floor? We hope Contraption NO.6, the "Coin-Operated Toothpaste Dispenser," will provide that wacky blend of fun and ingenuity that will inject some sparkle into even the gloomiest of mornings in the bathroom. At the very least, you can look at it, smile, and say to yourself: "I made that, and it works (sometimes)." The additional projects also use similar principles.

> **SUGGESTED COMPONENTS**
> notice board • foam board • assorted pins • thumb tacks • adhesive • weights • wire • wood • rigid plastic pipe • and four coins…

Rules and Levels

We've given Contraption NO.6 a difficulty rating of 4—because it has a few more moving parts, rather than actually being that much more difficult to put together than Contraption NO.1. It has seven different components, individually, they aren't that complicated, apart from the "Striking Pivot" device on the bottom left-hand side and the

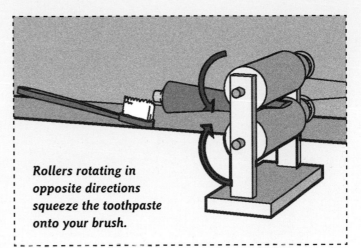

Rollers rotating in opposite directions squeeze the toothpaste onto your brush.

"Squeezer Rollers." If you follow our materials recommendations, you should be able to put the project together in under half a day. As a contraption, this reminded us of some old penny-in-the-slot fairground machines, before the advent of electronics, when the fun-value of the game depended on the maker's ingenuity and some very simple principles. This illustrates the rule of having an element of surprise in your contraptions perfectly. One coin would have been too easy, so four is much more enterprizing.

Bits and Pieces

What we have in mind is a very compact and totally portable design—maybe 3 ft (1 m) square in size. All the elements are made of some form of foam board and are fastened onto a notice board, which could be cork or another suitable material. In terms of the tool kit and materials needed, this is probably one of the easiest projects to make. Once it's finished, you might want to cover the whole thing with a sheet of clear plastic, to keep it dry and stop the coins from rolling off. The only parts that can't be made out of board are the rollers,

and we'd suggest sections of plastic waste pipe that are rigid and strong, yet light, but they could also be sections from a rolling pin or small bottles.

Power and Principles

There is a single power source driving the device: our old friend, gravity. If in doubt, drop it, roll it, or swing it. But within that, there is the application of several different principles. There are three levers—two obvious seesaws at the beginning, and a third in the pivot-device on the bottom left of the board. A pendulum fires the final stage that is operated by a weight-and-pulley arrangement that will provide enough mechanical advantage to squeeze the toothpaste out of the tube.

Variations on a Theme

Contraption NO.7 (see p. 42) preserves the original design, but adds a conveyor belt to move the toothbrush; Contraption NO.8 (see p. 43) keeps the conveyor, but changes the arrangement on the board; Contraption NO.9 (see p. 44) goes for a new outcome, while Contraption NO.10 (see p. 45) adds a twist to squeezing the toothpaste.

Coin-Operated Toothpaste Squeezer

This contraption consists mainly of flat elements that are fixed to a notice board in such a way that they can rotate and move. The challenge is to fit them all on the notice board and get them to work together. The process is started by a rolling coin at the top of the contraption, setting off a chain reaction that leads—eventually—to the release of a weight that pulls a string that then turns two "Squeezer Rollers" that squeeze the toothpaste tube between them.

 Level: *4*

 Time: *less than half a day*

1	Insert the coin on the top left. It drops onto a seesaw, which it tips, before carrying on down a further three inclined planes.

2	The first seesaw nudges a second, displacing coin number two.

3	Coin two falls onto another pivoting device that rotates and releases coin number three, which falls onto an inclined plane and comes to rest in a gap between two pieces of cardboard.

4	Meanwhile coin one comes to rest on the pivot-bar assembly, which drops and causes the vertical pivot bar to rotate, knocking coin three from its resting place.

5	Coin three then rolls down the slope and strikes coin number four, a pendulum, which swings into the catch holding the weight and releases it.

6	The catch could consist of a loop in the string hooked over a tack, and a small wedge that pushed the string away from the notice board when the pendulum strikes it.

7	An arrangement of two pulleys translates the force of the weight to act on the rollers that rotate to squeeze the toothpaste tube onto the waiting toothbrush.

CONTRAPTION CONNECTIONS

For the flat components of this contraption you can use thick cardboard, or foam board. The latter is super-light, versatile, tough, and easy to cut and shape; unfortunately, it's also probably the least eco-friendly material ever made—we'll leave the final choice to your conscience. The movement of rotating parts, such as the main pivoting device **A**, can be adjusted by tightening or loosening the tacks that hold them to the board. Make sure that the notice board itself **B** doesn't tilt forward at all, or the coins will simply fall off. The fixings for the pulleys **C** and **D** will need to be strong to support the weight **E**. The catch **F** must be strong enough to hold the weight, but delicate enough to release easily. The "Squeezer Roller" device **G** is connected to the board at one end, but it needs firm support on its free side, so make this part out of solid wood.

CONTRAPTION NO.6

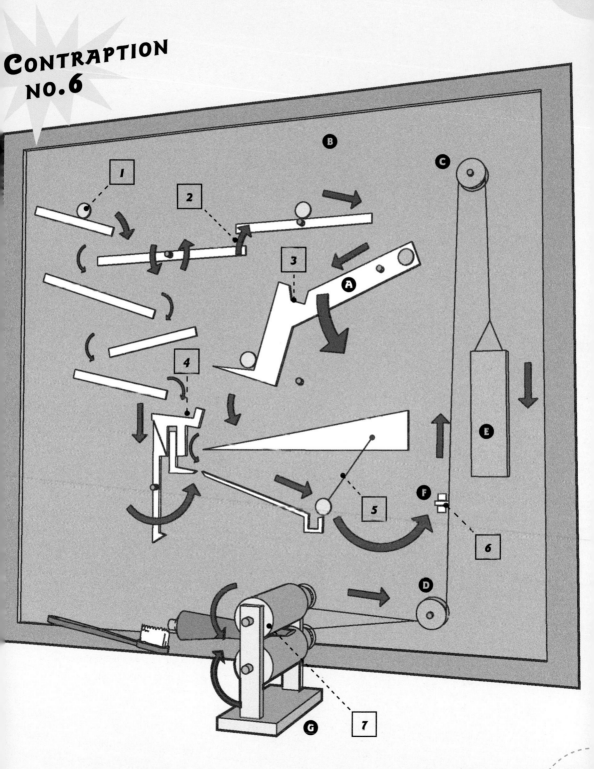

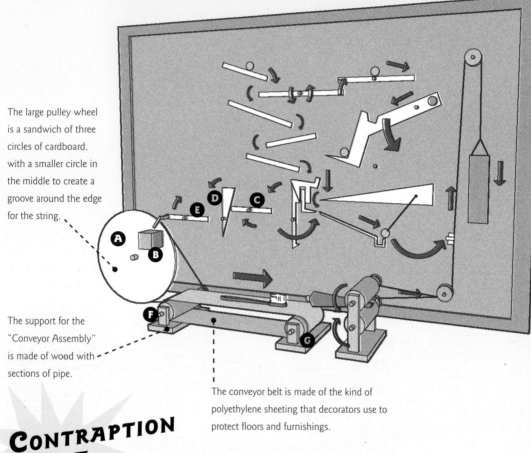

The large pulley wheel is a sandwich of three circles of cardboard, with a smaller circle in the middle to create a groove around the edge for the string.

The support for the "Conveyor Assembly" is made of wood with sections of pipe.

The conveyor belt is made of the kind of polyethylene sheeting that decorators use to protect floors and furnishings.

CONTRAPTION NO. 7

Push Me, Pull You

In this project, the upper part of the design remains mostly the same as before, but we've added a toothbrush conveyor to move the brush into position just as the toothpaste is extruded from the tube. Timing is paramount, or you'll be left with a mess. To unlock the wheel **A** powering the conveyor, and allow the attached weight **B** to turn it, we've added three more pivoted elements **C**, **D**, and **E**, on the board, which are triggered by coin one falling on the pivot-bar device. The "Conveyor Assembly" needs to be firm and stable, so heavier materials, such as wood for the frame and sections of metal tubing for the rollers **F** and **G** work best.

Level: 5

Time: one day

Level: 5

Time: one day

CONTRAPTION NO.8

Leaps of Faith

This time, we've changed the upper part of the contraption to give it a little bit of extra zing at the start. The coin takes a ski-jump **A** onto a seesaw that directs it along a curve, which propels it over a small chasm **B** onto the "Striking Pivot." (Why not see how wide you can make the gap?) This sends the coin down another curve **C**, along two inclined planes **D**, onto the pivot-arm device as before. The rest of the contraption—the motions of coins two to four, and the roller and conveyor elements—are as before. To adapt the previous project to this one would take less than an hour but, starting from scratch, we've given a total build-time of one day.

The combination of curves and pivoting elements gives this design some extra visual interest.

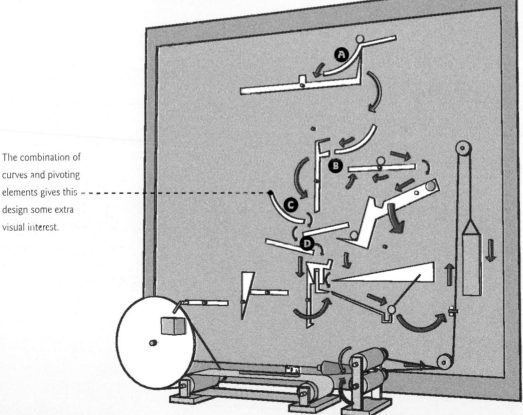

Level: *4*

Time: *less than a day*

CONTRAPTION NO.9

The Little Squirt

Instead of squeezing a toothpaste tube, this contraption delivers a squirt of soap, hand cream, or suntan lotion, depending on your choice of product, under a "Squirting Arm." We've kept the overall arrangement on the board, but we've lost the conveyor and roller mechanisms. We've replaced these with a seesaw **A**, one end of which is pulled up by the descending weight **B**, pressing the other end down on the container **C**. The seesaw needs to be made of a material that won't bend, such as wood or even a length of copper tubing. Fix the base down so the force presses down on the bottle, rather than lifting the seesaw!

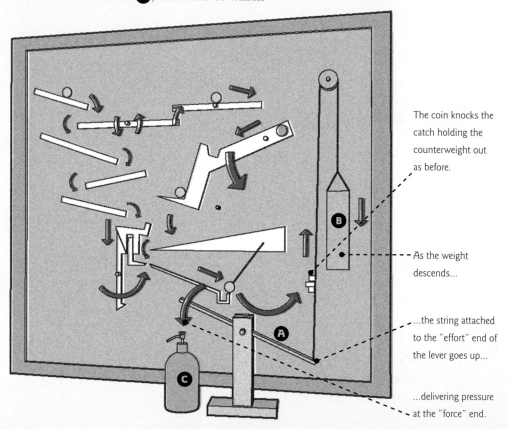

The coin knocks the catch holding the counterweight out as before.

As the weight descends...

...the string attached to the "effort" end of the lever goes up...

...delivering pressure at the "force" end.

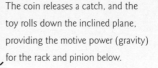

The coin releases a catch, and the toy rolls down the inclined plane, providing the motive power (gravity) for the rack and pinion below.

As the cog turns, two dowels roll the end of the tube, squeezing the toothpaste onto your waiting brush.

The neck of the tube is suspended in place with string.

Level: 7

Time: *one and a half days*

CONTRAPTION NO.10

Auto-Power

This contraption also squeezes toothpaste, but the power source for the "Conveyor Assembly" and "Squeezer Rollers" is a different application of gravity—an inclined plane **A** combined with gearing. In this case, the motive power is provided by the weight of a toy cart **B** pulling on a string. Even with the bulk of the contraption built, getting the rack **C** and pinion **D** to work smoothly together can be a tricky job. These two elements convert the linear motion of the toy cart into rotational motion to roll up the toothpaste tube between the two protruding dowel pegs. The gear assembly can be made of cardboard with matchstick teeth or, if you have the tools, it could be made of wood, which would undoubtedly be stronger.

CONTRAPTIONS 11–15

We all know that plants can only thrive with sunlight and water, but too much or too little of either is likely to be fatal. Water is easy to control as a houseplant is indoors, but sunlight isn't something you can order at will. That overcast morning when you left for work might turn into a scorching afternoon, leaving your houseplant dangerously exposed. Contraption engineers to the rescue!

Solar-Powered Plant Shader Projects

Unusually for a contraption, these projects combine the best principles of contraption engineering with a useful function that could be achieved by other means only with the greatest of difficulty or cost (for example, by hiring a private indoor gardener). Of course, the job could be done by a simplified device, but that's not what contraptions are about.

 Power: *radiant energy, gravity*

 Principles: *lever, projectile, wedge, pendulum, spring*

The Inspiration

Plant care isn't everyone's forte. Some of us mean instant death to any houseplant we're given, no matter how resistant. It's not that we don't like houseplants, but we have the knack of zeroing in on a plant's weaknesses, and making sure that it gets exactly what it doesn't need—in spades. If a plant is dry loving, we'll drown it; if it needs regular watering, we'll let it die of thirst; if it needs shade—you get the picture? We kill them with well-meaning but inappropriate kindness. Cue the "Solar-Powered Plant Shader," and at least one of the common causes of houseplant fatality—excess sunlight and heat—will be banished; and at a fraction of the cost that would be incurred by installing photochromic glass (glass that becomes more opaque as sunlight increases) in your windows or hiring a houseplant attendant to do the job for you. Of course, you do have to remember to prime the device itself; but, it's a start. Rome, as they say, wasn't built in a day.

SUGGESTED COMPONENTS
wood • strip of absorbent material • two coat hangers • golf ball • rubber ball • polystyrene cup • pencil • and a Japanese paper fan...

Rules and Levels

The five component parts of this project all need a certain amount of elaboration, so we've given this project a rating of 5 out of 10. However, even with this level of difficulty, we've estimated it shouldn't take you more than a day to build and get operational. We had a lot of fun with this, and we used the tried-and-tested principle of "Mixing It Up," using different power sources, principles, and materials. We also used a lot of recycled materials that we found lying around the house and garage. The intended task, needless to say, is also classically simple.

Bits and Pieces

Finding the best material to use for the "Shrinking Strip" will demand a certain amount of trial and error on your part. The intensity of the sunlight in Edinburgh, Scotland, will hardly compare with that in Houston, Texas, so how much absorbency the strip will need will vary with the climate at your location. A sheet of light cotton might do in cooler latitudes, while you might be going for an extra-thick material in warmer ones. The "Cannonball Run" projectile that shoots down the rails made out of coat-hanger wire is our old friend, a golf ball. The cup is of the standard-issue polystyrene variety, the "Battering Ram," a pencil, and the spring that will release its stored energy once the catch is released is a small strip of metal, wood, or plastic. The "Concertina Shader" itself is a decorated paper Japanese fan, which we found lying around, with a rubber ball attached to its end to act as a counterweight.

A fun way for those with or without green thumbs to look after their plants.

Power and Principles

We liked this device a lot, because once it's primed by soaking the "Shrinking Strip," it's completely automatic. You don't even have to be in the room, though it would be a shame to miss it the first time around. The initial motive force comes from the shrinking of a strip of material soaked in water, caused by the radiant energy of the sun. The shrinkage of the material tips a seesaw (first-class lever), pushing a ball (projectile) along a track to strike the cup mounted on an arm that rolls down a wedge. The cup strikes the suspended "Battering Ram"—a pendulum—that swings to release the stored energy in a very simple spring. Gravity provides the final motive force as a weight pulls the fan open.

Variations on a Theme

Contraption NO.12 (see p. 52) uses a variant of the initial solar-power device, and leaves the rest largely unchanged, apart from the shape of the projectile and the first ramming device; the back end of the contraption—the fan arrangement and its immediate power source—is what changes in Contraption NO.13 (see p. 53). In Contraptions NO.14 and NO.15 (see pp. 54 and 55), we've kept the strip idea, but this time we've experimented with some different shader options.

CONTRAPTION NO. 11

 Level: 5

 Time: a day

Solar-Powered Plant Shader

What makes this contraption appealing is its original and effective use of solar power. The soaking-wet strip of material that shrinks as it dries in the sun, is a bit of contraption-engineering magic, and is an innovative way of starting things off. As is the case for several of the devices in this book, the constituent components of the basic design are found objects, rather than things that have to be made, so this is a quick project to put together. It's also visually interesting, somewhat humorous, and ecologically friendly. We've specified a list of items on the previous page, but in this design, in particular, you've got a free hand to improvise with whatever materials you might have at hand. The principle is simple, so feel free to do it your own way.

1 Once the device is primed by soaking the cloth strip, it's up to the sun to do the rest.

2 One of our favorites—a golf ball—is tilted off the seesaw onto the rails.

3 The rails, made out of coat-hanger wire, lead the ball into the waiting polystyrene cup.

4 An inclined plane helps the cup to achieve the right velocity.

5 The cup impacts the pencil pendulum that rams into the catch holding the spring that keeps the fan shut.

6 The spring gives it the initial push and the counterweight does the rest.

7 A rubber ball acts as a counterweight to the fan.

G

6

F

5

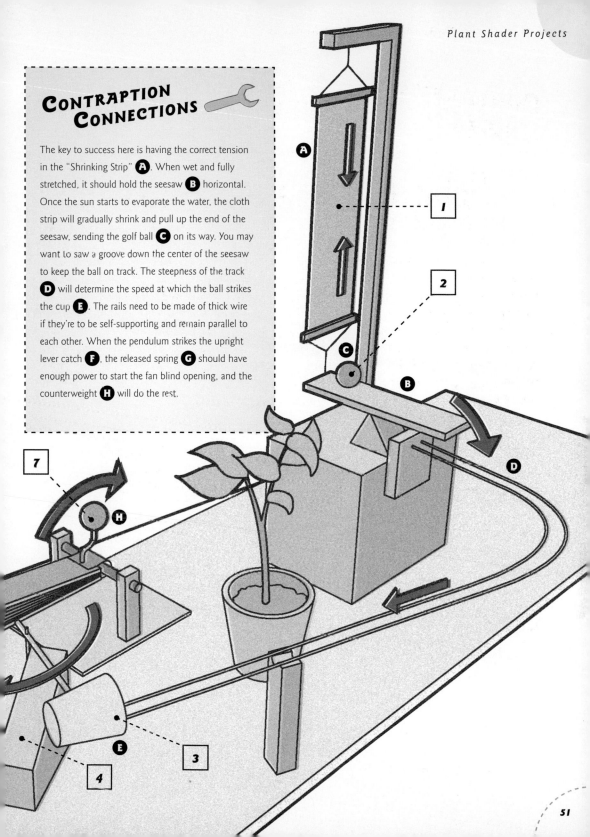

CONTRAPTION CONNECTIONS

The key to success here is having the correct tension in the "Shrinking Strip" **A**. When wet and fully stretched, it should hold the seesaw **B** horizontal. Once the sun starts to evaporate the water, the cloth strip will gradually shrink and pull up the end of the seesaw, sending the golf ball **C** on its way. You may want to saw a groove down the center of the seesaw to keep the ball on track. The steepness of the track **D** will determine the speed at which the ball strikes the cup **E**. The rails need to be made of thick wire if they're to be self-supporting and remain parallel to each other. When the pendulum strikes the upright lever catch **F**, the released spring **G** should have enough power to start the fan blind opening, and the counterweight **H** will do the rest.

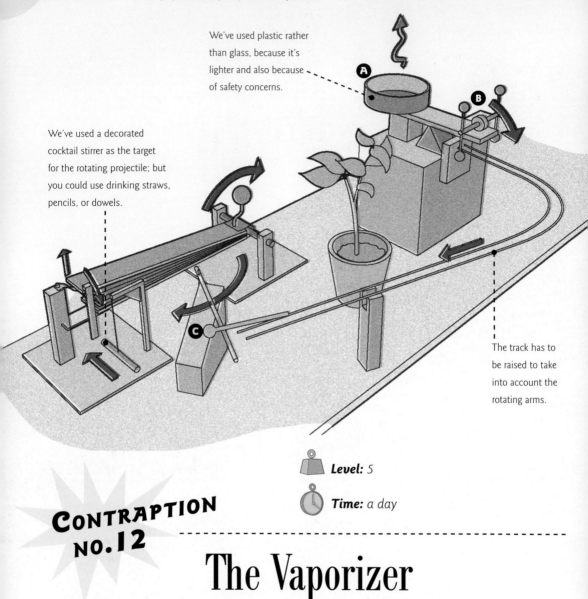

We've used plastic rather than glass, because it's lighter and also because of safety concerns.

We've used a decorated cocktail stirrer as the target for the rotating projectile; but you could use drinking straws, pencils, or dowels.

The track has to be raised to take into account the rotating arms.

Level: 5

Time: a day

CONTRAPTION NO.12

The Vaporizer

The sun's radiant heat—the basis of solar power and of all life on earth—is a very versatile resource. This second application also uses heat and evaporation as its initial trigger. This time the water in a small plastic container **A** evaporates, allowing the lever to tip and the rotator **B** (a central wheel and axle with two sets of counterweights) to roll off the seesaw and down the track. The counterweights, which could be ball bearings or small fishing weights, will spin rapidly as the rotator travels along the rails. The new projectile design demands a slightly different-shaped target **C**.

CONTRAPTION NO. 13

Level: 6

Time: *a day*

The Concertina

The front end of the contraption—strip, projectile, rails, and cup—remains the same in this project, but this time the shade assembly is totally different. First, the cup hits a wooden wedge **A** resting on rollers. These could be sections of a broom handle, or you might use toy wheels instead. The forward movement of the wedge lifts a stiff flap **B** that acts as a catch, and this allows two large elastic bands under tension **C** to contract and pull the "Concertina Shade" **D** open along a horizontal wooden track.

With the far end fastened to the base of the strip, and the other to the horizontal strip, the shade will open smoothly and stay in place.

The greater the stretch in the elastic, the faster the shade will open, when the elastic snaps back to its original length.

To avoid tearing the paper blind, make sure it reaches the stop on the strip before it's fully stretched.

The rollers could be cardboard or plastic candy containers. Get something rigid that won't flatten.

The Venetian Rollerblind

 Level: 5

 Time: *a day*

This contraption may look complicated, but it isn't really. The function remains the same—a shade hides the plant—but we've also added a watering device. It'll require the cutting of some components, almost all of which can be made from wood. The "Shrinking Strip" is now longer, and when it shrinks it will exert a downward pull on a double-pulley arrangement **A** and **B** that pulls the top slat **C** up, tipping a rod onto another two slats. This disturbs two pairs of linked balls **D** and **E** that hold the blind in place. The descending balls (small, solid-rubber balls will work well) release the blind, and a three-pulley arrangement uses the energy of the descending blind to tilt the water bottle **F** and water the plant.

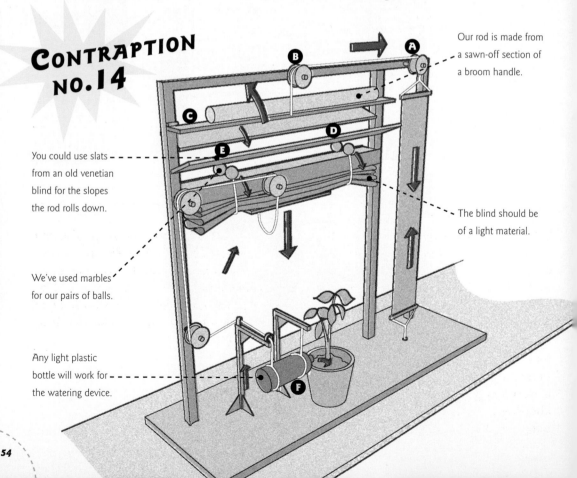

CONTRAPTION NO. 14

Our rod is made from a sawn-off section of a broom handle.

You could use slats from an old venetian blind for the slopes the rod rolls down.

We've used marbles for our pairs of balls.

The blind should be of a light material.

Any light plastic bottle will work for the watering device.

CONTRAPTION NO. 15

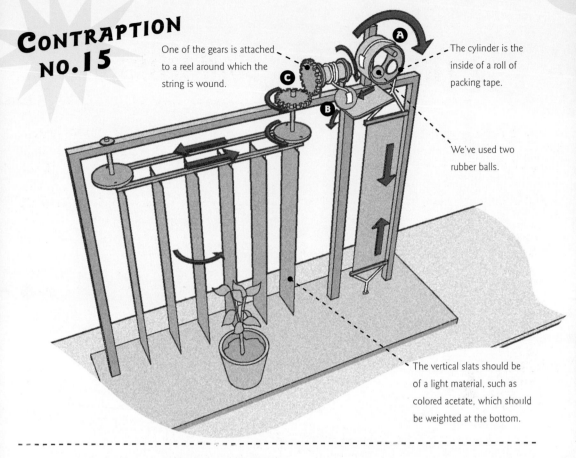

One of the gears is attached to a reel around which the string is wound.

The cylinder is the inside of a roll of packing tape.

We've used two rubber balls.

The vertical slats should be of a light material, such as colored acetate, which should be weighted at the bottom.

The Metropolis

This project is reminiscent of the infernal machines in Fritz Lang's nightmare vision of the future, *Metropolis* (1927). Made in Rube Goldberg's lifetime, it presented a far more sinister view of humanity's mechanized destiny than his light-hearted cartoons. Apart from the cogs, which are a bit tricky, this is probably the easiest project in this section. The "Shrinking Strip" turns a cylinder **A** with a hole cut out of its side, allowing a small ball to fall out and dislodge the large ball **B**, which is attached to a piece of string. The weight of the ball turns the gear mechanism **C** that rotates the screens until they shut. You can cut the cogs out whole from cardboard or wood, or you can cut out circles and stick the teeth on afterward (using dowels or matchsticks, depending on size).

 Level: 5

 Time: *less than a day*

CONTRAPTIONS 16–20

Water has long been used for its cooling, soothing, and decorative effects, and fountains have adorned the gardens and courtyards of Old-World palaces since antiquity. In olden days, before the advent of mechanical pumps, water engineers achieved the most extraordinary effects with the application of the simple principles of gravity, differential pressure, and hydraulic displacement.

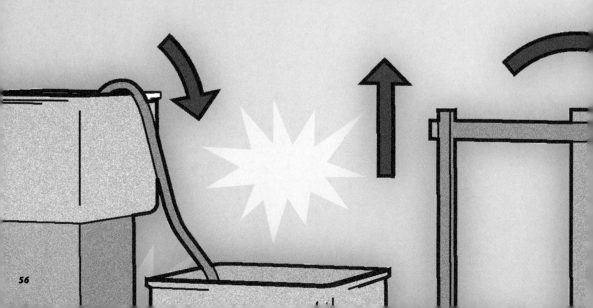

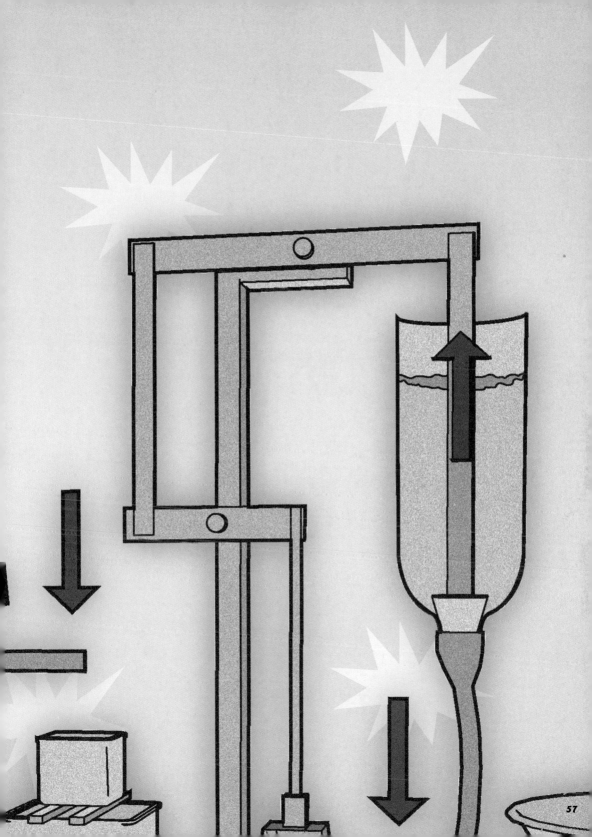

Tabletop Fountain Projects

Before the advent of modern technologies, one of the main sources of motive power available was good old H$_2$O. From the water wheels of the ancients to the steam engines of the Industrial Revolution, water has made the world what it is today. In these projects we make full use of the earth's most abundant and cheapest resource.

The Inspiration

The gentle play of water has long been recognized as a calming influence. The Arab peoples use fountains to cool and beautify their homes, and the Chinese assert that a strategically placed water feature in your living room will do no end of good for your feng shui. But the inspiration for this piece—believe it or not—are the

This water fountain is best used for decoration, rather than drinking from.

 Power: *gravity, air pressure*

 Principles: *lever, suction, displacement*

grandiose garden fountains of the palace of Versailles, home of the "Sun King," Louis XIV (1638–1715). The fountains of Versailles are fed and operated by some 125 miles (200 km) of tunnels, canals, channels, and aqueducts, which were originally powered by horsepower (of the animal variety), windmills, and gravity. Our own version of this wonder of French hydraulic engineering, however, will fit comfortably on your kitchen table.

SUGGESTED COMPONENTS

assorted plastic containers • balloon • plastic bottle • plastic tube • modeling putty • block or weight • two bricks • and lots and lots of water...

Rules and Levels

With its five components, four of which need little elaboration, a "Tabletop Fountain" should take you less than half a day to assemble. We've given it a difficulty rating of 3 out of 10, making it a good one to start with as a novice. In terms of the rules of contraption engineering, we haven't gone for variety here. On the contrary, we made full use of as many of the properties of one element as possible.

Bits and Pieces

We've gone for different sizes and shapes of plastic containers, but if you're careful, you could go for glass containers and drinking glasses. The fountain bowl itself, however, will have to be plastic so that you can make a hole for the pipe in the base. We've kept it fresh and gone for a round pastic bowl. The pipe inlet is sealed with modeling putty, but you could also use waterproof adhesive or bathroom sealant. The hardest component, from a manufacturing point of view, will be the "Lever-Operated Plug," and you might want to try an easier solution, like a pulley, for example.

Power and Principles

The fountain kicks off with a "Vacuum Siphon," which is an application of suction—the creation of a partial vacuum or area of low pressure in the tube. Once you've created the vacuum by sucking out the air, the siphon will function on its own. Remember that the end of the tube in the bottom container must be lower than the surface of the liquid in the upper container for the siphon to work. The weight of the water compresses the air in the balloon and tilts the "Air Scales" (a first-class lever), plunging the weight or block into the water

EUREKA!

Everyone knows about Archimedes jumping out of the bathtub and running down the street without so much as a towel, shouting "Eureka!" (Ancient Greek: "I've got it!"). But can you remember what he was so excited about? Supposedly, Archimedes was hired by the king of Syracuse to discover whether his new crown was made of pure gold, or whether the jeweler had cheated him and adulterated it with silver. It would have been easy if he'd been able to melt it down, but he had to do it in such a way that the crown remained in one piece. It was in his bath that he realized that a body immersed in water displaced a certain volume. Dividing the mass of the crown by its volume would give the crown's density—this is now known as the "Archimedes' Principle." Already knowing the densities of silver and gold, Archimedes was then able to tell if the crown was pure gold.

("Eureka Device"). The water is displaced (see box) and overflows. Gravity and leverage pull out the cork, and the pressure forces the water entering the bowl into fountain.

Variations on a Theme

Contraption NO.17 (see p. 62) gives a whole new "cool" outcome; Contraptions NO.18 (see p. 63) and NO.19 (see p. 64) add a "sweet" new beginning to the basic design; while NO.20 (see p. 65) starts with a bang!

Tabletop Fountain

This water feature on your kitchen table is a lot of fun to assemble, and you won't need expensive components. We've gone for a selection of containers, with a bowl for the fountain basin itself; the water tower that feeds the fountain is an upside-down plastic bottle. The other parts can be built from off-cuts of timber. This contraption demonstrates several of the principles of hydraulic engineering including siphoning, displacement, and pressure.

Level: *3*

Time: *less than half a day*

1 Sucking on the bottom of the tube will set the water moving from one container to the other. The tube's end must be lower than the liquid's surface in the top container.

2 The weight of the water pushes the air from one side of the balloon to the other, raising one end of the seesaw.

3 Don't overfill the balloon; it'll need to be malleable, and not so full that it bursts.

4 The block lowers and displaces water that overflows into the bottom container: a "Eureka Device!"

5 The extra weight tips a second seesaw and operates the three-lever assembly that uncorks the bottle.

6 The differential pressure forces the water to become a fountain. Experiment with nozzle shapes and sizes to get different spray heights and patterns.

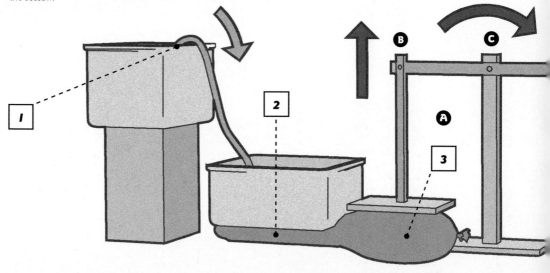

CONTRAPTION NO.16

CONTRAPTION CONNECTIONS

Start by constructing the balloon balance shown at **A**. This is a simple wood construction with bolt fastenings at points **B** and **C** to allow for the flexing of the joints to swing when the balloon rises. The "Water Tower" is constructed in a similar manner, making sure that the base plate **D** has a bottom large enough to handle the weight of the water without toppling over. The bottle itself needs to be securely supported on a separate unit or upright surface to allow for the stopper to be removed when the arm **E** swings up. When attaching the tube to the bottle spout **F** immerse it in hot water to make it more pliable and easier to squeeze over the bottle neck. Ensure that the seal is good. Before adding water to any of the containers, be sure to have a dry run—testing will help you to avoid a soaking.

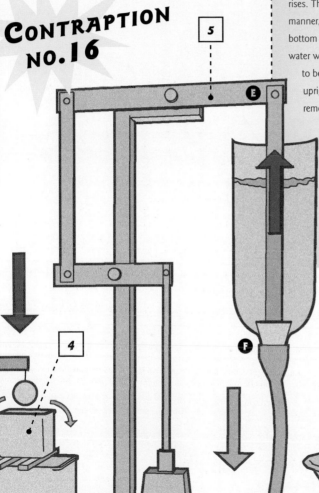

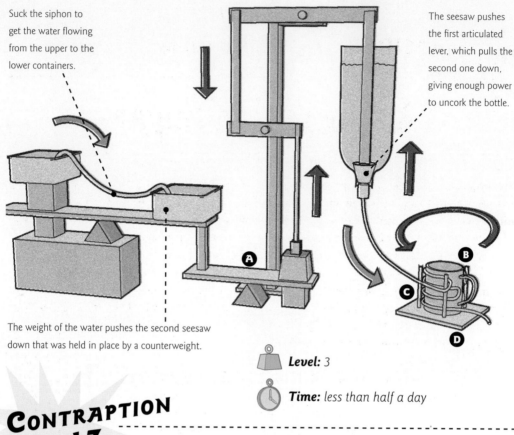

Suck the siphon to get the water flowing from the upper to the lower containers.

The seesaw pushes the first articulated lever, which pulls the second one down, giving enough power to uncork the bottle.

The weight of the water pushes the second seesaw down that was held in place by a counterweight.

Level: 3

Time: less than half a day

CONTRAPTION NO.17

Water Cooler

This contraption replaces the fountain with a hot-beverage cooling mechanism, taking the heat out of your coffee or tea by surrounding the cup in a cool-water enclosure. It starts off as before with a "Vacuum Siphon," with the second container balanced on the end of a seesaw (that is, a first-class lever). Gone is the "Eureka Device," which is replaced by a second lever device **A** that will operate the de-corking assembly. In this instance, a weight powers two articulated levers that transmit the force. Instead of the fountain, we have the water-cooler mechanism itself, which is made of a thin plastic tube wound around four dowelling posts to create a cooling cradle for a cup containing the hot beverage of your choice. Reusing the components from the "Tabletop Fountain," simply place four wooden dowels or old pens **B** in a firm base **D**, such as florist's foam or polystyrene, and weave the thin tubing **C** around the frame to create the cooling cradle.

Level: 3

Time: half a day

CONTRAPTION NO. 18

A Spoonful of Sugar

This project makes use of another property of water: that it provides the medium in which certain solids can dissolve, as their molecules become evenly dispersed in the fluid. When the "Sugar-Cube Tower" **A** dissolves, the weighted lever **B** tips back, pulling the plug (a simple bath plug will do) from the second container, which empties itself into another bowl **C**. The weight needs to be enough to pull the plug,

and the "Sugar-Cube Tower" will support a small rock. Depending on the size of the hole you make in the second container, you can obtain a trickle or a waterfall. The weight of the water filling the lower container displaces the air in a semi-inflated balloon, which raises a second seesaw **D**, which is the "Eureka Device" used earlier.

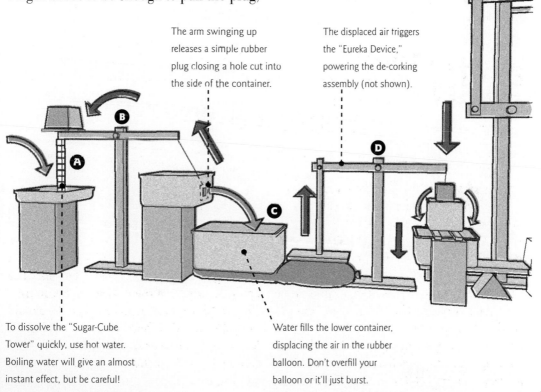

The arm swinging up releases a simple rubber plug closing a hole cut into the side of the container.

The displaced air triggers the "Eureka Device," powering the de-corking assembly (not shown).

To dissolve the "Sugar-Cube Tower" quickly, use hot water. Boiling water will give an almost instant effect, but be careful!

Water fills the lower container, displacing the air in the rubber balloon. Don't overfill your balloon or it'll just burst.

CONTRAPTION NO. 19

Level: *4*

Time: *less than a day*

Sugar Tower

This contraption introduces a completely new front end into the design. It starts with a siphon as before, but this time, filling of the lower container pushes up a simple ball-and-rod assembly. The rod displaces a larger ball, the weight of which tips a bowl of water into the container holding the "Sugar-Cube Tower." The frame at **A** can be easily constructed with garden canes in a wig-wam fashion with a pulley used at **B**. String, wool, or clothesline can be used at **C**. Use a section of drain pipe at **D** and polystyrene or light wood for the rod with a ping-pong ball at the base. Be sure to check that the weight shown at **E** isn't in-line with the swinging ball when it's dislodged from the pipe.

The larger ball needs to be fairly heavy, like a baseball, to tip up the water container.

We've cut a notch on one side of the container to make it easier for it to empty and create the waterfall effect.

The small ball needs to have good buoyancy, and we've used a ping-pong ball.

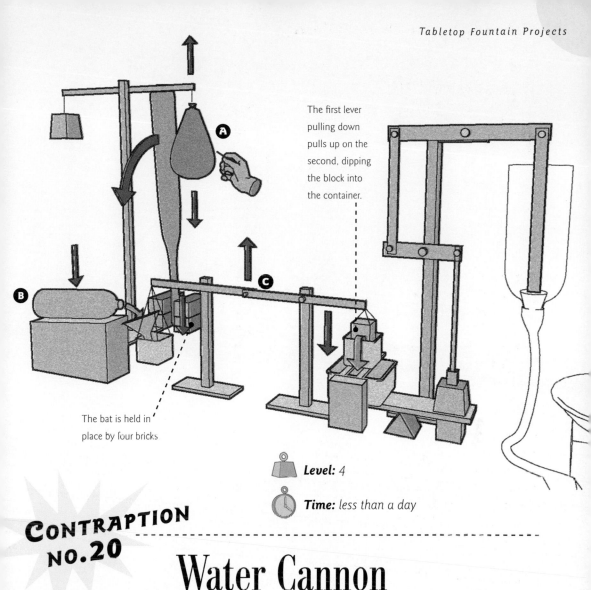

The first lever pulling down pulls up on the second, dipping the block into the container.

The bat is held in place by four bricks

Level: *4*

Time: *less than a day*

CONTRAPTION NO.20

Water Cannon

This is our most "explosive" contraption to date, with a lot of action in its first three stages. Keep children, pets, and those of a nervous disposition well clear when setting it off! Bursting the water-filled balloon **A** allows the weight at the other end to tilt the lever, releasing the baseball bat. The bat falls onto the plastic bottle **B**, or "Water Cannon," sending a jet of water into the funnel and into the container, triggering the double-lever arrangement **C**, and this operates the "Eureka Device," and so on to the fountain. Place blocks (for example, you could use house bricks) on each side of the bat handle to ensure that it falls in a straight line.

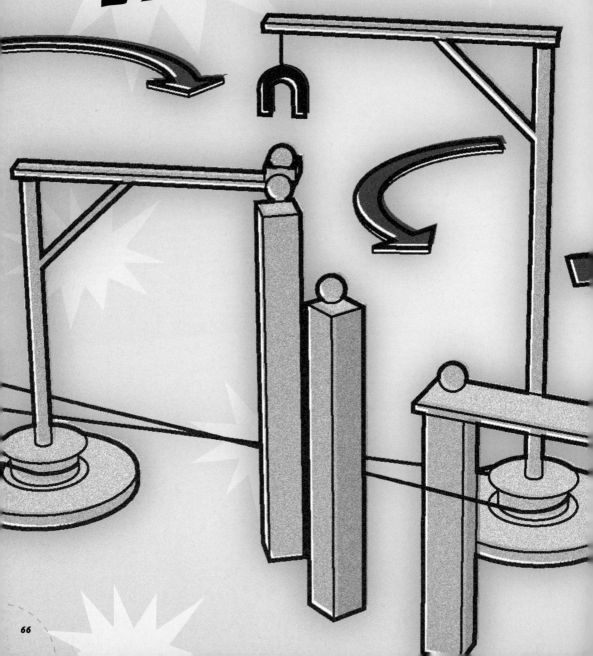

Getting object A into receptacle B is a favorite task of contraption engineers. In these projects, we've combined mind-boggling, awe-inspiring complexity with utility by making a group of devices that adds a touch of sweetness to the hot beverage of your choice.

Magneto Sugar-Cube Dispenser Projects

These projects are a prime example of a real contraption classic. The task—sweetening a beverage—is simplicity itself, and a something that every one of us has done a thousand times, so the challenge here was to make it as complicated as possible.

 Power: *air pressure, gravity, magnetism*

 Principles: *air brake, pulley, gears, wheel and axle, lever, pendulum*

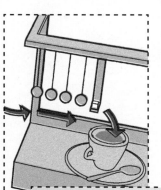

A simple "Newton's Cradle," which is effectively a series of small pendulums; the last one of which knocks the sugar cube into the waiting cup.

Suggested Components

wood • string • cardboard • weight • five cotton spools • magnet • nine ball bearings • and a sugar cube...

The Inspiration

Authors are notoriously addicted to tea and coffee—it must be sitting at a desk all day with only a computer screen for company. So it should come as no surprise that a contraption mysteriously materialized between point A, the sugar bowl, and point B, the coffee cup. The challenge then was to see how many stages, devices, and principles we could employ to get from A to B. Breakfast-related devices have long been a favorite of contraption engineers. Rube Goldberg and Heath Robinson both came up with their own versions, and both Nick Park's *Wallace and Gromit* series and the *Back to the Future* movies feature contraptions that butter toast, fry eggs, and brew coffee and tea.

Rules and Levels

The principle this contraption illustrates to perfection is complexity ad absurdam (which is Latin for "This is totally insane"). Find the easiest, most mundane everyday task—something that wouldn't even register on your scale of noticeable things—and put in as many stages as possible to accomplish it. There are a couple of tricky parts to this contraption: the cog assembly, the timing of the crane, and the "Newton's Cradle," so we've given it a 7 out of 10, with a build-time of one day.

Bits and Pieces

Because the components of the device need to be fairly sturdy, we've made most of it out of wood. The sails of the "Air Brake" can be made of cardboard or paper, depending on how much braking power you need, which will depend on the weight you've used. The cogs, as in other examples we've used previously, can be cut out of wood in one piece, or you can use matchsticks for the individual teeth. Cotton spools make good pulleys and wheel-and-axle devices, and we've used them in the "Air Brake," the cogs, and in the two cranes. Ball bearings come in a variety of sizes and alloys, but the basic steel kind will do fine for the four balls the "Magneto Crane" has to pick up, and for the "Newton's Cradle."

Power and Principles

The windmill at the start don't power the device, but instead acts as an "Air Brake" to slow the descent of the weight. The motive power that sets off the device, therefore, is our old friend gravity. The weight acts through a pulley to turn a gear assembly of three cogs that translates the force from vertical to horizontal, and also splits it into two. The cranes rotate in opposite directions on their wheel-and-axle bases. The magnet attracts the steel ball bearings, temporarily magnetizing them. The last bearing balances a lever, which tilts as the supporting weight is removed.

STUCK ON YOU

Magnets, or loadstones, (see pp. 22–23) have always fascinated humans for their strange power of attracting or repelling objects. All objects are affected by magnetic fields but metals, especially iron, steel, and nickel, are particularly susceptible to magnetism. James Clerk Maxwell (1831–1879) described the basic laws of magnetism; but these were not fully understood until Albert Einstein (1879–1955) produced his Theory of Relativity. The movement of electrically charged particles such as electrons causes the phenomena of magnetism and electromagnetism. In the latter, those excitable electrons create an electric current, which powers our lights, computers, televisions, and household appliances. Permanent magnets, however, are created by the quantum-mechanical spin and orbital motion of electrons within the atoms that make up magnetic ores and minerals.

Variations on a Theme

In Contraptions NO.22 and NO.23 (see pp. 72–73), we've kept the sugar-dispensing function but altered the mechanisms and principles involved in its delivery to the cup; Contraption NO.24 (see p. 74) substitutes stirring for sweetening; and NO.25 (see p. 75) keeps the front-end mechanism but gives you a whole new outcome.

Magneto Sugar-Cube Dispenser

This invaluable device uses the energy of a falling weight to swing two cranes, one of which picks up the ball bearings with a magnet until it finally releases the first ball in a "Newton's Cradle" to knock a sugar lump into the waiting cup. This project requires fairly precise construction, so if you're not feeling confident, it might be best to develop your contraptioneering skills by practicing on simpler projects first.

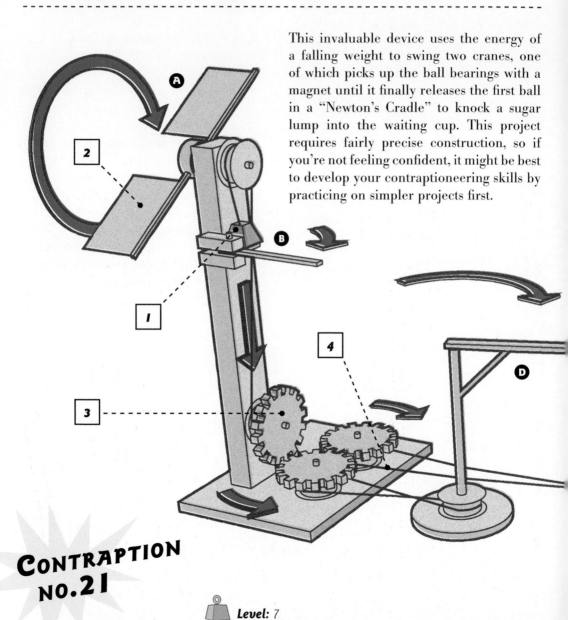

CONTRAPTION NO.21

Level: 7

Time: one day

CONTRAPTION CONNECTIONS

Although it looks like a source of power, the windmill **A** is only here to slow the rate at which the falling weight **B**, the real source of energy, descends. The sails need to be light and stiff, and sheet plastic, glued into slits of thin dowel, would work just as well as cardboard or paper. So that the "Magneto Crane" **C** picks up the first ball bearing, adjust the starting position of the second crane **D** so that they meet. The only materials you might have to buy are the magnet itself, which needs to have a strong pull, and the ball bearings, although you could obtain small ones by dismantling the hub of an old bicycle wheel. Skills-set and tool-wise, you'll be using your woodworker's hat and goggles for this contraption.

1 Releasing the catch allows the weight to drop, as gravity drags it downward.

2 The windmill sails act as an air brake to slow down the descent of the weight.

3 The descending weight turns the first cog that then powers two other cogs.

4 The cogs translate the movement to the cranes and split the force along two pathways.

5 The cranes rotate in opposite directions, causing the "Magneto Crane" to pick up the four ball bearings.

6 The seesaw tips releasing the last ball bearing that forms part of a "Newton's Cradle."

7 The force of the first ball bearing is transmitted to the next, and this continues along the line, until the last one knocks the sugar cube into the cup.

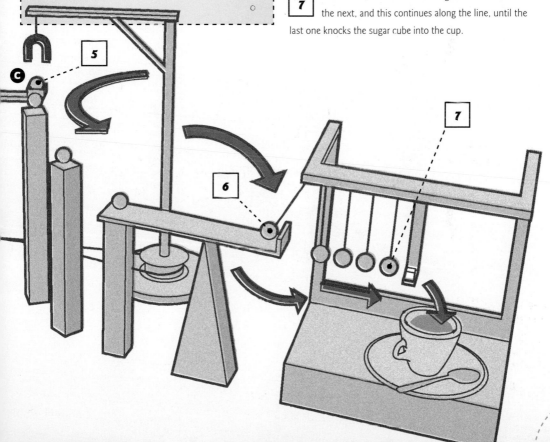

Wagon Wheels

 Level: 7

 Time: *one day*

This project retains the front and back ends of the previous contraption, but replaces the cranes and ball bearings with two wheels. These wheels and their pulleys need to be rigid, so if you use cardboard it should be fairly thick. The supports should be made of wood. The cog assembly rotates the first wheel **A** until the ball bearing **B** rolls through the hole in the wheel, onto the gantry **C**, and into the box on wheel two **D**. The string on this second wheel has been left loose so that it'll turn only once the ball bearing is in position. Carried by the second wheel, the ball bearing rolls off, knocks another one **E** along the seesaw and activates the "Newton's Cradle."

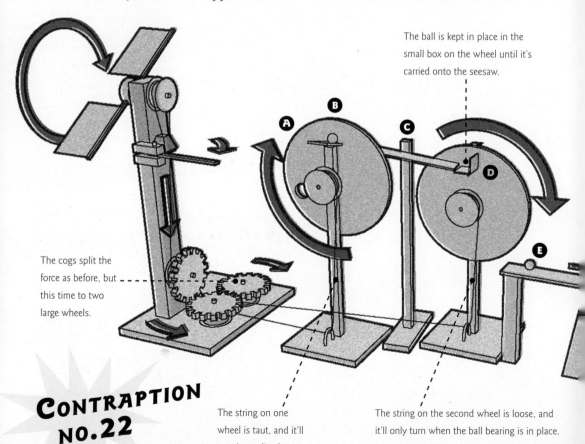

The ball is kept in place in the small box on the wheel until it's carried onto the seesaw.

The cogs split the force as before, but this time to two large wheels.

The string on one wheel is taut, and it'll turn immediately.

The string on the second wheel is loose, and it'll only turn when the ball bearing is in place.

CONTRAPTION NO.22

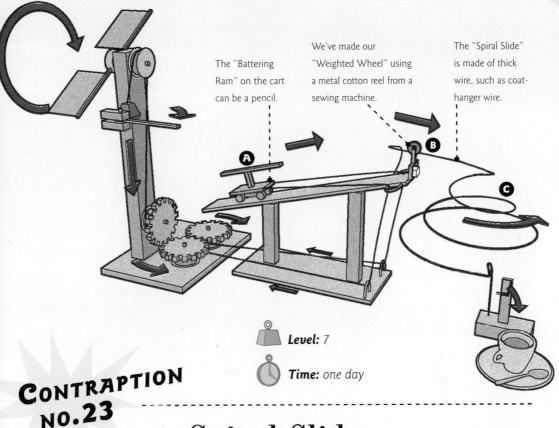

The "Battering Ram" on the cart can be a pencil.

We've made our "Weighted Wheel" using a metal cotton reel from a sewing machine.

The "Spiral Slide" is made of thick wire, such as coat-hanger wire.

Level: 7

Time: one day

CONTRAPTION NO.23

Spiral Slide

Again, the front end remains the same, but the cogs now tow a cart **A** up an inclined plane, which can be made of cardboard, providing it's stable. The battering ram on the cart hits the "Weighted Wheel" **B** at the top of the "Spiral Slide" **C**. This could be a metal spool or even the wheel from a clothesline pulley. The wheel is kept in place on the track by a weight that hangs below it. The wheel spirals down the track (which should be made of stiff wire that is attached firmly at both ends), gaining speed as it goes, and knocks the sugar cube into the waiting cup.

CONTRAPTION NO.24

Level: *6*

Time: *less than a day*

Stirred not Shaken

What would have James Bond have given for this ingenious device to stir his tea? The lever **A** releases the catch holding the weight and powers the cogs. This time the force is split via a pulley arrangement. One cog drives a turntable **B** on which the cup rests, rotating it clockwise, while the other turns a wheel and axle that rotate a spoon **C** counterclockwise, making for a perfectly stirred cup of British tea. The pulleys on the vertical rods can be cotton reels, and the bottom turntable could be a tabletop "Lazy Susan" with the string looped around its rim. Make sure that the two main parts of the contraption are fixed firmly in relation to each other, to keep the right tension on the pulleys.

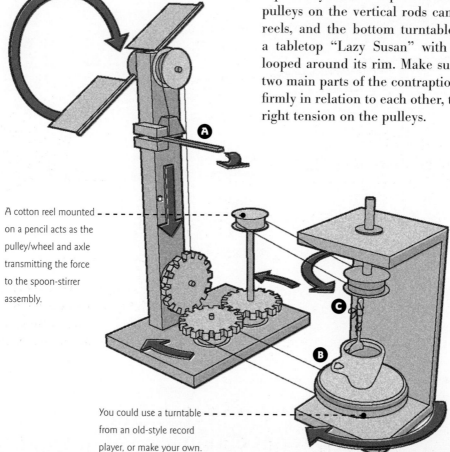

A cotton reel mounted on a pencil acts as the pulley/wheel and axle transmitting the force to the spoon-stirrer assembly.

You could use a turntable from an old-style record player, or make your own.

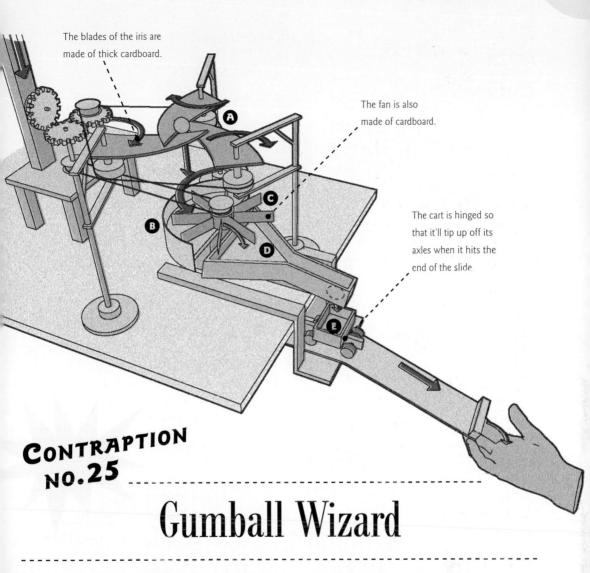

The blades of the iris are made of thick cardboard.

A

The fan is also made of cardboard.

C

B

D

The cart is hinged so that it'll tip up off its axles when it hits the end of the slide.

E

CONTRAPTION NO.25

Gumball Wizard

Here, the cogs drive two different mechanisms. The "Gumball Iris" consists of three blades **A**, made of plastic or cardboard, that support a gumball between them. As the cogs turn, they rotate the wooden rods to which the blades are fixed, opening the iris and allowing the ball to fall into the semicircular tray **B** below. A revolving fan **C** then pushes the ball off the tray into a flat funnel **D**, and it falls through a hole into the waiting cart **E**. The cart (plywood will work better than cardboard) rolls down to the stop at the bottom. It then flips up and deposits the gumball into your waiting hand.

 Level: 7

 Time: one day

CONTRAPTIONS 26–30

ere's a clever idea for the security-conscious snake lover. You may want to let your snake out for its daily slither, but you don't want just anyone opening the tank allowing your precious reptile to roam at will. Alternatively, you may want to allow someone else to let your pet out, but also give them time to get out of the room.

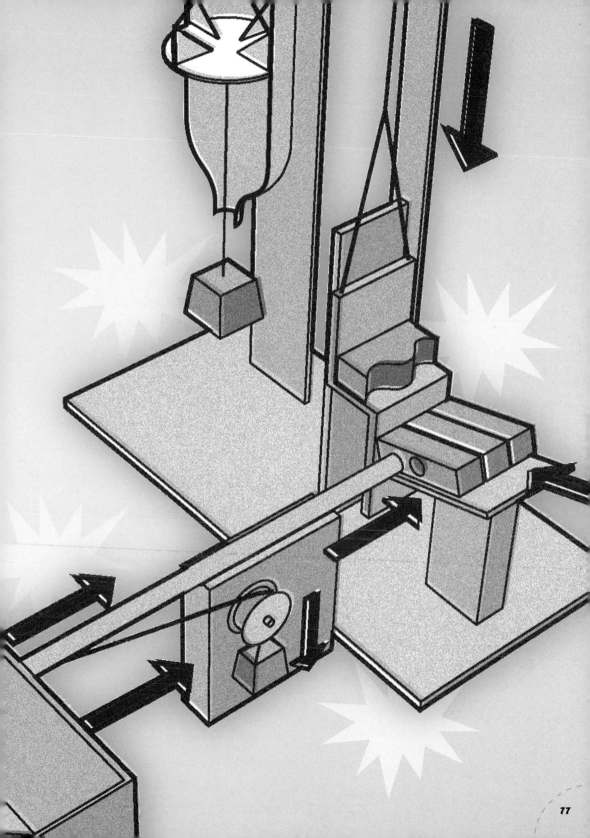

High-Security Snake Releaser Projects

Letting the pet out for its daily exercise can be a chore, so we've decided to inject a little bit of fun and fantasy into this routine task. Another issue we took into consideration was security: you don't just want to let anyone release your pet snake, scorpion, or tarantula, so these contraptions work with a key.

 Power: *gravity, spring*

 Principles: *air pressure, friction, counterweight, pulley, inclined plane*

The Inspiration

Imagine you've gone away on vacation and you've left your pet snake, Wanda, in the care of your sister-in-law, Janice. The trouble is, Wanda and Janice don't exactly hit it off. Janice is phobic about snakes, and Wanda loves to snuggle up. We've found a way to allow Janice—and only Janice—to open Wanda's tank, while remaining at a safe distance—in Janice's case, down the hall, locked in the bathroom. Once the mechanism is engaged, Janice has plenty of time to hightail it to safety. (Please note: no snakes were harmed in the making of this contraption.) If you don't have a pet snake, or a phobic sister-in-law, then these projects are a fun way to guard your secret diary, or something equally precious.

Rules and Levels

We had to put a fair amount of thought into this contraption, so the contraptioneering principle it demonstrates is ingenuity. The front- and back-end mechanisms are fairly straightforward, but the lock mechanism in the middle has to be quite carefully aligned, or it won't function, even with the correct key. However, we've rated this project as a 5 out of 10, with a build-time of a little more than a day.

SUGGESTED COMPONENTS
wood • plastic bottle • cardboard • string • three cotton spools • two weights • spring • box • hinge • and a rubber snake...

The stored energy in the spring enables the lid to flip open once the catch is released.

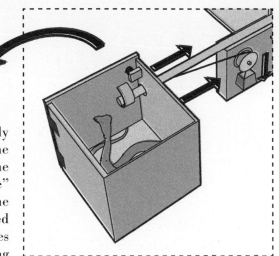

Bits and Pieces

The structural elements have to be fairly sturdy, so we've gone for wood for all the supports, the lock mechanism, and, the snake tank, of course. The "Bottle Brake" is not a power unit, but acts to slow the descent of the shelf on which you've placed the "Security Key." Once the key reaches the three blocks of the "Block Lock" resting on a slope, they slide forward, and because you've used the right key, the holes bored through the three sections of the lock will align. The rod shoots through the hole and releases the hinged "Spring-loaded Door" on the tank.

Open and Shut Case

Look around your home and the hinge is probably the most common device you've never really noticed. Hinges are everywhere, on doors, of course, cabinets, kitchen appliances, laptops, toilet seats, and CD players. The first recorded hinges date back to the second millennium BCE, to the cities of the ancient Near East. Early hinges were simple wooden pivots set in stone sockets (imagine a wheel with a long axle holding the door). Our modern metal door hinges were invented in the nineteenth century.

Power and Principles

The initial motive power is provided by gravity directed through a pulley. But, to slow the device down, we've installed the "Bottle Brake," which uses the friction of the cardboard against the side of the bottle, a counterweight, and the differential air pressure caused by the small bottle-neck through which the air enters the bottle. A third pulley/gravity arrangement powers the rod that slides through the "Block Lock." The final power component is provided by a spring held in place by a catch. The release of the stored force in the spring pushes the "Spring-Loaded Door" open.

Variations on a Theme

Contraption NO.27 (see p. 82) retains the front and middle mechanisms, but changes the outcome. In Contraptions NO.28 (see p. 83), NO.29 (see p. 84), and NO.30 (see p. 85), we return to the snake-release theme, but change the ways in which the contraption is put together.

High-Security Snake Releaser

Rube Goldberg and Heath Robinson often introduced animals into their designs, but as active components in the mechanism and not just as passive observers, as in this instance. In our own animal-rights conscious age, having a golf ball hit a penguin or using a cat on a treadmill as a power source clearly isn't right. If you are going to involve your pets in your contraptions, make sure that nothing can go wrong and injure or scare them.

 Level: 5

 Time: *more than a day*

1 The bottle is shown here in cutaway, but that is just to show you the internal workings. The bottle's only opening should be at the bottom and the rest should be as airtight as possible.

2 The key is placed on the shelf, and the extra weight causes it to descend.

3 The slow descent enables the person opening the tank to leave the room.

4 The blocks slide down the inclined plane into the descending key. Make sure that there is something stopping them being pushed backward off the shelf.

5 Only the correctly shaped "Security Key" will allow the holes in the "Block Lock" to align correctly. Make several different shaped keys to keep the device secure.

6 Once the blocks are all aligned, the rod shoots through the hole, pulled by the weight.

7 As the rod withdraws from the box, it frees the catch, releasing the "Spring-Loaded Door."

Contraption Connections

This contraption needs to be fairly sturdy, and it has some large areas of material, so particle board or oriented strand board may be easier to use than wood. The door of the box containing the snake is spring-loaded **A**, but is held shut by the rod **B**. The rod, made of dowel, needs to move out of the catch, so the spring shouldn't be so strong that it grips the dowel and prevents it from sliding. Even the small spring from a press-top ballpoint pen will be enough to push the door open, and you can hinge the door at the bottom if this makes opening it easier. Make sure that the bottoms and sides of the three wooden blocks **C**, and the surface on which they rest, are completely smooth, so that they can slide easily down the slope. You might use sheet metal for the slope, and adjust the angle of it as necessary.

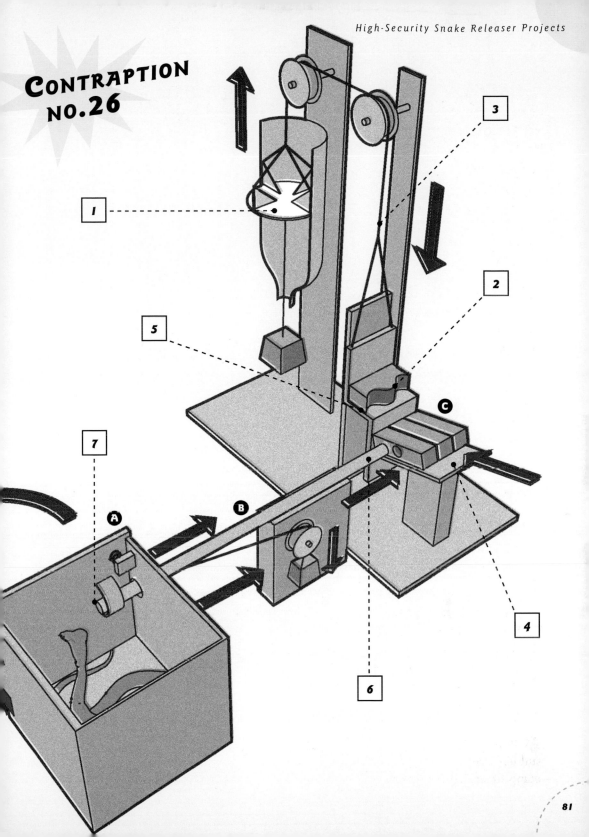

CONTRAPTION NO.26

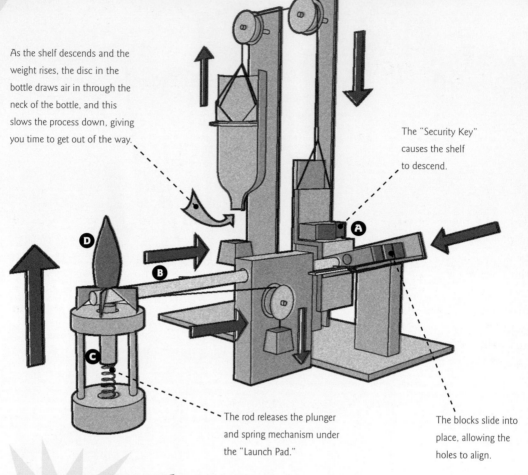

As the shelf descends and the weight rises, the disc in the bottle draws air in through the neck of the bottle, and this slows the process down, giving you time to get out of the way.

The "Security Key" causes the shelf to descend.

The rod releases the plunger and spring mechanism under the "Launch Pad."

The blocks slide into place, allowing the holes to align.

CONTRAPTION NO.27

Liftoff

While the front end remains unchanged in this project, we've lost the snake and replaced it with a rocket instead. When the "Security Key" **A** falls into place, the blocks align, and the rod **B** slides into the holes. This releases the spring-loaded plunger **C** that powers the "Launch Pad." The sudden release of energy from the compressed spring fires the rocket **D**.

The plunger can be a piece of dowel, and a valve spring from a engine is a perfect power source, providing it isn't under so much pressure that the rod can't slide out.

 Level: 5

 Time: more than a day

CONTRAPTION NO. 28

Level: 5

Time: *more than a day*

Sliding Doors

This is a rather elegant contraption. The initial part of the contraption works as before, but the opening mechanism for the doors of the box is totally new. In this instance, the rod is attached to four pieces of string **A** that are in turn attached to two light "Sliding Doors" **B** and **C**. The doors are held shut by a string with three small weights on it **D**. When the rod retracts, they're pulled around the columns **E** and **F**, and back into the sides of the box. As the doors open, the weights are drawn apart until they end in a straight line. The lighter and smoother the doors, the better they'll operate, so use shiny thin plastic if possible.

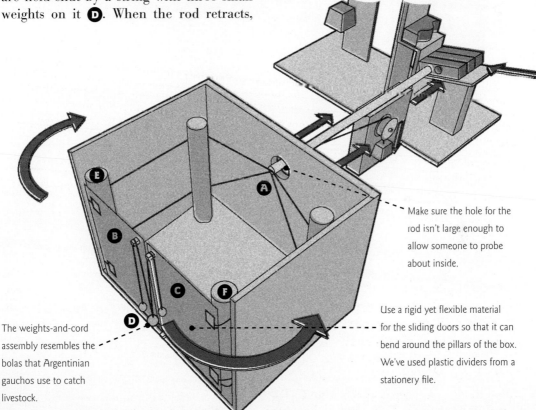

Make sure the hole for the rod isn't large enough to allow someone to probe about inside.

The weights-and-cord assembly resembles the bolas that Argentinian gauchos use to catch livestock.

Use a rigid yet flexible material for the sliding doors so that it can bend around the pillars of the box. We've used plastic dividers from a stationery file.

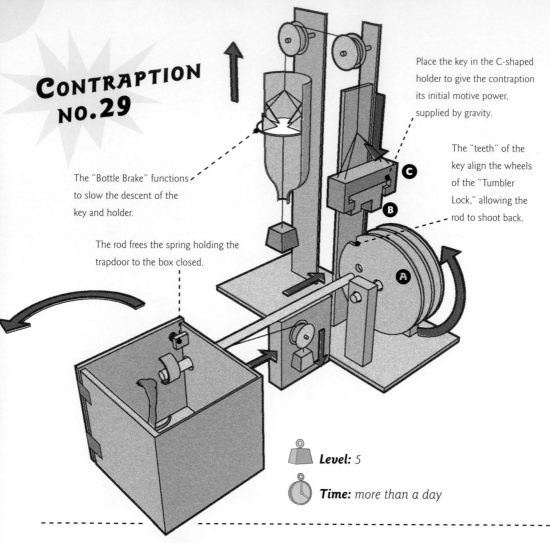

Place the key in the C-shaped holder to give the contraption its initial motive power, supplied by gravity.

The "teeth" of the key align the wheels of the "Tumbler Lock," allowing the rod to shoot back.

The "Bottle Brake" functions to slow the descent of the key and holder.

The rod frees the spring holding the trapdoor to the box closed.

C

B

A

Level: 5

Time: *more than a day*

The Tumbler Lock

In the "Tumbler Lock," we've altered the lock-and-key mechanism. Instead of three blocks, the key aligns three rotating disks **A**—a simplified version of a standard Yale lock. The disks can be made from thick cardboard. You start the contraption by slotting the key **B** into a C-shaped holder **C**, which starts to move downward. The "Bottle Brake" slows it down as before. Each disk has a notch cut out of its edge that fits the key. When the key slots in, it pushes the disks around, aligning the holes in them, and the rod shoots through. The calibration of the "Tumbler Lock" might take you a bit of time, so we've given this project a build-time of more than a day.

Level: 5

Time: one day

CONTRAPTION NO.30

The Trawler

In this instance, the principle that slows the key's descent is based on the greater resistance of water over air. The "Trawler Brake" mechanism consists of a shallow tank of water **A** with a plate **B** at one end. The tank could be a plastic container, providing its sides match the profile of the plate. The plate can be made from wood or thick plastic that won't bend as its pulled through the water. The plate is attached to the shelf by a string and pulley **C**. As the shelf descends, it's forced to drag the plate through the water. The rest of the contraption works as before. As the "Block Lock" opens the rod withdraws, releasing the door catch and allowing the spring to push the door open.

The mechanism slowing the descent of the key is a plate of wood drawn through a container of water.

Place the key on the ledge to start the contraption.

The "Block Lock" falls into place and the rod shoots back.

The "Spring-Loaded Door" opens, allowing your pet to slither away happily.

CONTRAPTIONS
31-35

Unlike the projects so far, the projects in this section are literally a pure flight of fancy. With one "crushing" exception, they serve no useful purpose—no simple task made absurdly complicated—other than the sheer pleasure of demonstrating the best principles of contraption engineering.

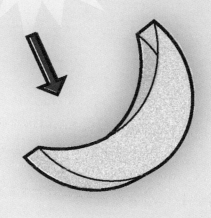

Wind-Powered Rocket Launcher Projects

"Fly me to the moon and let me play among the stars...." Johnny Mathis sang many years ago. Three of these contraptions give you all the excitement and anticipation of a moon shot in your living room.
5... 4... 3... 2... 1... BLAST OFF!

Power: *air pressure, elastic band*

Principles: *wheel and axle, pulley, projectile*

These projects use wind power, which is converted by a variety of components into the energy stored in a stretched elastic band, which launches your rocket in a graceful arc.

The Inspiration

As children, there were some things that fascinated us and that, even when we've grown up, we can never quite let go of. Let's face it, we're all big kids at heart, especially when it comes to things that go bang, and those that are fired high into the air. Get something that goes bang and gets fired into the air, and you're onto a real winner. In the "Wind-Powered Rocket Launcher" we've gone for the second, but there's nothing to stop you from adding something a little explosive (like a bursting balloon) to coincide with the launch mechanism.

SUGGESTED COMPONENTS
wood • cotton spool • cardboard • string • elastic bands • and a toy rocket...

Rules and Levels

This was a fun contraption to design, and we hope it'll be a fun one to build. Unlike the previous projects, it serves no useful function, however impractically realized. The principles that this contraption best exemplifies are the elements of surprise and suspense. Surprise because the function of the contraption may not be immediately obvious, and suspense because, depending on the strength of the wind, the ratchet may take some time to reach the launch position. The ratchet-and-latch mechanism may be a bit tricky to calibrate, but overall we've given this project a 6 out of 10, with a build-time of less than a day.

Bits and Pieces

The wheel and axle—made of a large cotton reel, on which the windmill sails are attached—are mounted on a sturdy wooden stand. The sails need to be fairly rigid, so we've used cardboard mounted on thin wooden strips, rather than paper. The windmill is linked to the ratchet mechanism by a pulley, and as it turns, it pulls the ratchet down. The ratchet stretches an elastic band as it descends. When it's almost all the way down, the ratchet pushes in a rod that is connected to an oval stop. With the stop removed, the latch plate drops and slides away from the latch. This frees the ratchet, which shoots upward propelled by elastic. The rocket resting on the "Launch Pad" attached to the ratchet is fired upward into the waiting "Moon Target."

Power and Principles

Wind power is a familiar concept now that the wind turbines that generate electricity are part of our landscapes. However, histor-ically this is the second major application of the windmill principle, as from the twelfth to the nineteenth centuries windmills (along with watermills) were a very useful power source (see below). In this project, we've employed a windmill with a pulley as the main power source. The ratchet-and-latch mechanism draws down the "Launch Pad," which is attached to the second power unit, the elastic band (see p. 22–23). The rocket, once launched, becomes a projectile, which would elegantly demonstrate Newton's Law of Universal Gravitation, if it weren't for the waiting moon.

Blown Away

The first windmills were made in Persia in the ninth century, but the windmills we're familiar with were invented in Europe in the twelfth century. Originally, windmills were constructed in such a way that they could be turned to face into the wind. At first, the whole building rotated and had to be of light construction, but later, only the topmost part, or cap, turned while the rest of the windmill remained stationary. This meant that they could be built of stone or brick, and they could be fitted with larger sails, allowing them to generate much more power. Windmills were used to grind corn and also to power sawmills, threshing machines, and water pumps to drain marshland and irrigate crops.

Variations on a Theme

In Contraption NO.32 (see p. 92), we've added an interesting refinement to the "Moon Target." In NOS.33–34 (see pp. 93–94) we've kept the rocket-launcher function, but we've gone for a completely different outcome in NO.35 (see p. 95).

CONTRAPTION NO.31

Level: 6

Time: *less than a day*

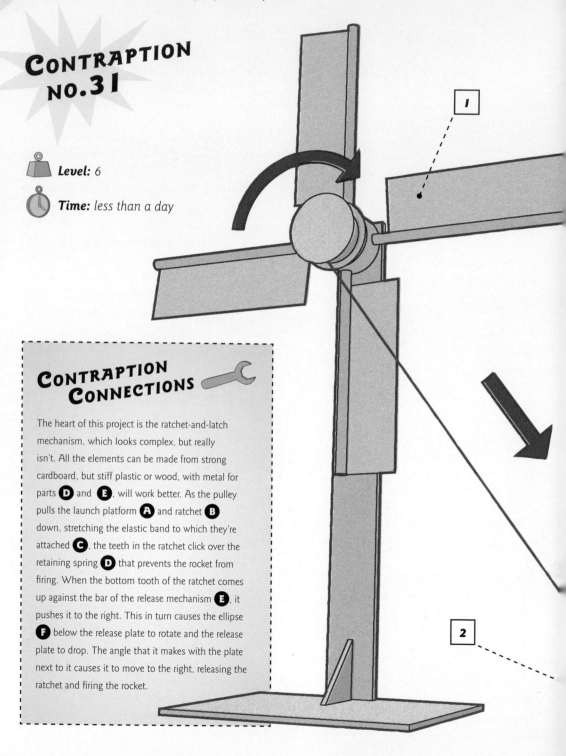

CONTRAPTION CONNECTIONS

The heart of this project is the ratchet-and-latch mechanism, which looks complex, but really isn't. All the elements can be made from strong cardboard, but stiff plastic or wood, with metal for parts **D** and **E**, will work better. As the pulley pulls the launch platform **A** and ratchet **B** down, stretching the elastic band to which they're attached **C**, the teeth in the ratchet click over the retaining spring **D** that prevents the rocket from firing. When the bottom tooth of the ratchet comes up against the bar of the release mechanism **E**, it pushes it to the right. This in turn causes the ellipse **F** below the release plate to rotate and the release plate to drop. The angle that it makes with the plate next to it causes it to move to the right, releasing the ratchet and firing the rocket.

Wind-Powered Rocket Launcher

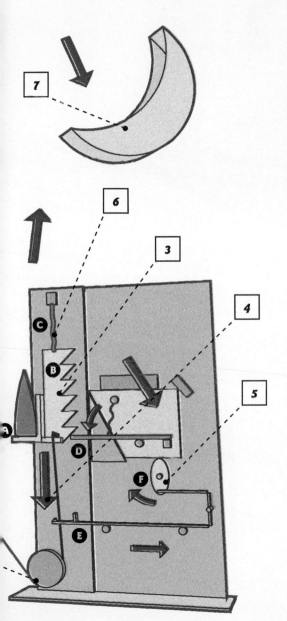

This is an unusual project as it really mixes up power sources and principles in an interesting way—and it's a fun and fairly unexpected outcome as well. The distance the rocket will travel will depend on the length and thickness of the elastic band you attach to the ratchet and the weight of your rocket, so you'll have to experiment with the height and position of the moon. As with all projectile contraptions, be careful to clear the decks before liftoff.

1 Open a window or turn on an electric fan to power the windmill.

2 The rotational force of the windmill is converted to a downward force by the pulley.

3 The ratchet mechanism draws down the elastic band and prevents premature release.

4 As the ratchet descends, it triggers its own release mechanism.

5 When the oval stop rotates, it allows the latch plate to slide down.

6 The stretched elastic band is released, and the rocket is launched.

7 The "Moon Target," which is a hollow shell made from cardboard, is positioned to receive the rocket.

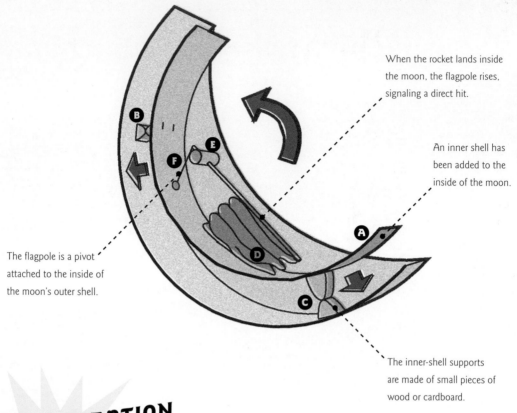

When the rocket lands inside the moon, the flagpole rises, signaling a direct hit.

An inner shell has been added to the inside of the moon.

The flagpole is a pivot attached to the inside of the moon's outer shell.

The inner-shell supports are made of small pieces of wood or cardboard.

CONTRAPTION NO.32

Stars and Stripes

This is a special project to create a more advanced moon target to go with your rocket launcher. Inside the moon (shown here in cross-section), there is an inner shell **A** made of cardboard delicately resting on two supports **B** and **C** on the outer shell. A furled flag **D** with a pivoted flagpole **E** is attached to the side of the outer shell, and the bottom end of it is attached to the inner shell by a thread **F**. When the rocket lands on the moon, the inner shell falls off its supports, dropping into the outer shell, pulling the thread, and hoisting the flag. This works best with light materials, so use a piece of thin nylon for the flag and balsa wood for the flagpole and pivot.

 Level: 6

 Time: *less than half a day*

CONTRAPTION NO.33

The Archimedes

This project has a new mechanism, based on an "Archimedes Screw." In this version of the design, a length of clear plastic tube is wound around a wooden rod . The rod is tied to the center of a wheel with string, so that it can be fixed at an angle without the need for a complicated joint. When the windmill turns the wheel, the rod rotates and pumps the water from the container up the tube, into a chute , and into the second container. Two hooked arms extend over the sides of the launch platform. The arms must be stiff, so use wood and attach them loosely so they rotate. The water in the second container pulls it down until a wooden plate attached to the support **C** pushes the arms outward and releases the launch pad.

Level: *7*

Time: *less than a day*

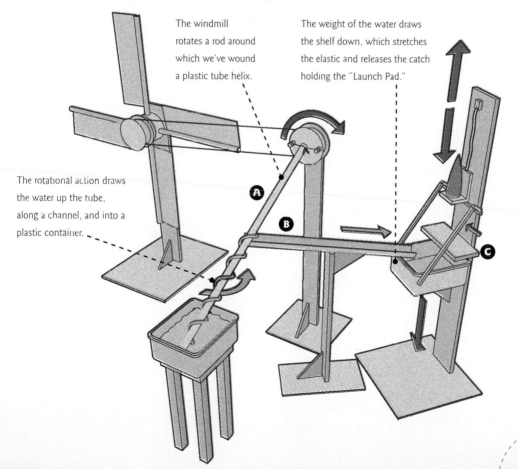

The windmill rotates a rod around which we've wound a plastic tube helix.

The weight of the water draws the shelf down, which stretches the elastic and releases the catch holding the "Launch Pad."

The rotational action draws the water up the tube, along a channel, and into a plastic container.

CONTRAPTION NO.34

 Level: 6

 Time: one day

Pulling Advantage

This contraption is all about power. The windmill is about half the size of the earlier one on pp. 90–91 because we've added a pulley mechanism **A** attached to the bottom of the ratchet. In fact, this is a textbook demonstration of the mechanical advantage (see pp. 20–21) that pulleys can create. The pulleys themselves can be made from wood, or you could reuse the wheels (without the tires) of a toy cart.

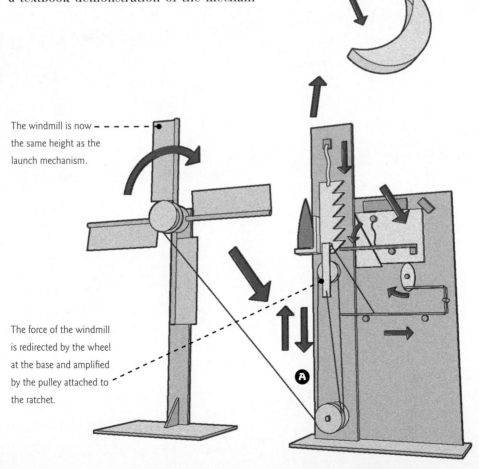

The windmill is now the same height as the launch mechanism.

The force of the windmill is redirected by the wheel at the base and amplified by the pulley attached to the ratchet.

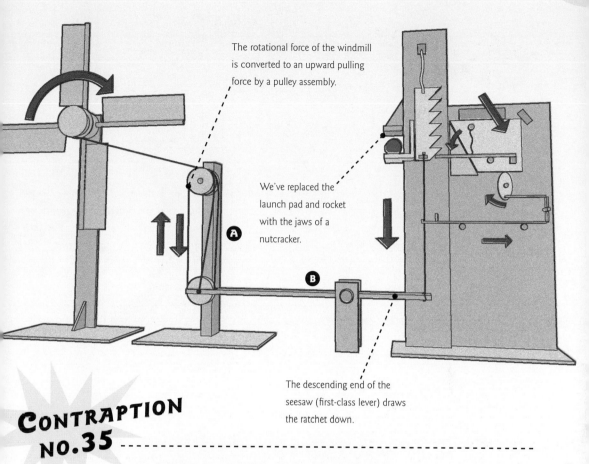

The rotational force of the windmill is converted to an upward pulling force by a pulley assembly.

We've replaced the launch pad and rocket with the jaws of a nutcracker.

A

B

The descending end of the seesaw (first-class lever) draws the ratchet down.

CONTRAPTION NO.35

The Nutcracker

As in the previous project, we've added a pulley **A**, but this time it is separate from the launch mechanism and is attached to a first-class lever (with the fulcrum between the force and the effort). The lever **B** can be made of wood, with a nail or dowel as the fulcrum. When the windmill turns, the pulley exerts a force at one end of the lever and pulls it upward. The force is transferred by the fulcrum to pull down the load at the other end, which is attached to

the ratchet by a piece of string. But wait a minute, where is the rocket? The "Launch Pad" on which the rocket rested is now the lower jaw of a nutcracker. Crunch! Use a hard material, such as wood or metal, for both the upper and lower jaws.

 Level: *6*

 Time: *one day*

CONTRAPTIONS 36–40

A larms are useful things: We use them to get up, to time our breakfast eggs, to turn on lights that scare off burglars, to remind us to do a hundred and one things day in, day out. So, why, oh why has no one ever designed an alarm that tells you when your bathtub is full? You need look no further than this group of ingenious contraptions with a bathtub theme.

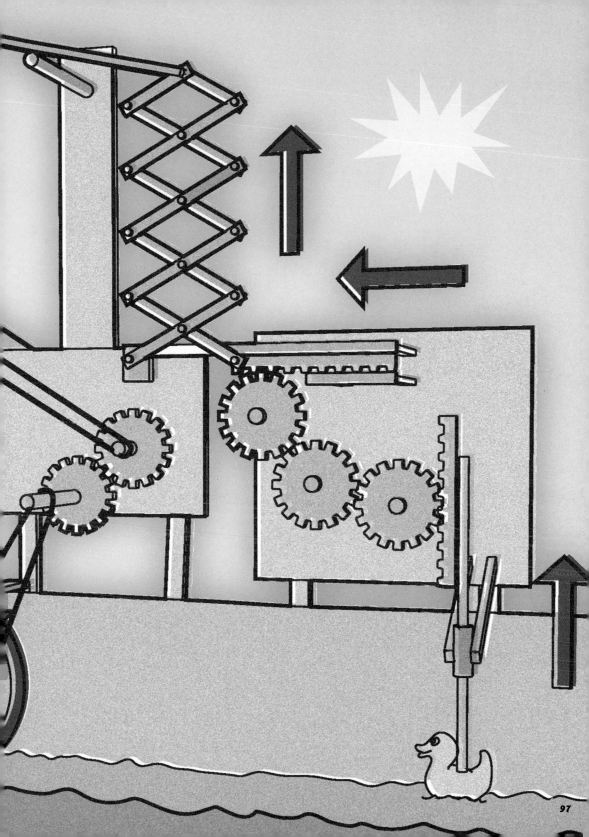

Water-Powered Bath Alarm Projects

The trouble with baths is that you turn on the faucet, and then get sidetracked: the telephone rings, your favorite TV program comes on, the kids come home from school—you know how it goes. But until you can afford a servant to run your bath for you, you'd be well advised to start building one of these ingenious devices.

 Power: *water pressure*

 Principles: *wheel and axle, ratchet and gears, lever*

The Inspiration

Imagine you're the top floor of your palatial Miami condo (once you've imagined that you own a palatial Miami condo, that is). There's nothing worse for your insurance company than when you're running a bath, you get engrossed in a half-an-hour phone call with your mother, wife, husband, son,

or daughter, and end up flooding out your three floors of priceless antique furniture, paintings, and rugs, as well as your neighbors below. If the insurance business knew about this contraption, they'd probably want it installed as standard in all bathrooms.

Rules and Levels

We've been saving this project for close to the end of the book because it's a tough one. You have to get everything exactly in the right place and make sure that the timing is just right, so we've given this 8 out of 10, with a build-time of a day. This project may be fairly elaborate, but it's a very pleasing one when you've sorted out the teething problems (no pun intended, though there are a lot of gears to adjust), and if we had to make up a word to describe the principle of contraption engineering that it illustrates, it would be "pleasingness"—the joy of a job well done.

> **SUGGESTED COMPONENTS**
> wood • plastic sheeting • nylon cord • concertina trivet • bell • and a rubber duck…

GOING WITH THE FLOW

A watermill is basically a turbine that converts the energy of flowing water into rotary motion. There are two types of turbine in common use: the reaction turbine, which changes the pressure of the water as it moves past the turbine blades; and the impulse turbine, which changes the velocity of the water flowing through the turbine. The technology has been in use since antiquity, and is most commonly used today for a renewable power.

Can you ring my bell? A novel use of water power.

The study of fluids comes under the science of "fluid mechanics," which is further divided into "fluid statics" (the behavior of fluids at rest) and "fluid dynamics" (the study of fluids in motion). Fluids obey the Newtonian laws of classical mechanics, such as the conservation of mass and momentum. Fluids such as water are known as "Newtonian fluids" because their viscosity only depends on temperature and pressure. A "non-Newtonian fluid," on the other hand, can become more or less viscous when a force is applied, for example, when banging the bottom of a bottle of ketchup.

Bits and Pieces

We've built the dry-land supports for this contraption out of wood, but anything that is likely to get wet should be made of plastic, such as the paddles of the "Faucet Watermill." We've also used nylon cord rather than regular string that could stick and rot. The cogs and racks are cut out of one piece, or you can stick matchsticks to a disk to make the teeth. The "Concertina Elevator" is an old metal trivet we found in the attic, but you'll also find concertinas holding up shaving mirrors. The duck is a standard-issue rubber model. As it floats upward, the rack-and-gear mechanism extends the concertina and lowers the "Alarm Bell" into the path of the second "Chiming Wheel" powered by the faucet.

Power and Principles

Water pressure provides the motive power for the "Faucet Watermill," which is a very simple turbine (see box), mounted on a wheel and axle. Water also powers the second half of the contraption by making the rubber duck rise. Both halves of the contraption employ gearing as their basic principle, to reverse the direction of the rotation on the right, and to raise the "Concertina Elevator" that is attached to a simple lever on the left.

Variations on a Theme

Contraption NO.37 (see p. 102) lifts the lid, quite literally, on a very different outcome. While Contraptions NO.38 (see p. 103) and NO.39 (see p. 104) retain the bath-alarm function, but change the alarm mechanism; and NO.40 (see p. 105) makes sure your bath is stirred and not shaken.

CONTRAPTION NO.36

Level: *8*

Time: *one day*

Water-Powered Bath Alarm

Flowing water and rising water combine in this contraption to let you know when your bath is filled to the right level, and ready for you to plunge in. A rubber duck—the one you always play with in the bath—acts as the float and motive power in a system of pulleys and levers that lowers a bell into the path of a rotating "Chiming Wheel," driven by a paddle wheel under the flow of water from the faucet and a sequence of pulleys and cogs, that strikes the bell. Of course, you could just stick around to keep an eye on your bath in case it overflows, but how much more fun would it be to hear the alarm bell ring as a result of your own handiwork?

CONTRAPTION CONNECTIONS

If at all possible, use waterproof materials for any of the elements that are likely to get wet. This applies especially to the "Faucet Watermill" and the spindle on which it turns **A**, which will be soaked. If you use plastic for the whole contraption, it'll increase the build-time, so use wood for the upper parts unless you can buy or recycle plastic parts for the racks and cogs. The elevator **B** can be made from popsicle sticks or wooden slats held together loosely with small nuts and bolts. If the wall is tiled, you could fix the contraption in place using double-sided tape.

1 The flow of the water turns the "Faucet Watermill" counterclockwise.

2 The blades of the watermill are made of a stiff plastic, as cardboard or wood would become waterlogged.

3 Thin nylon cord works better than regular string in this steamy environment.

4 Gears reverse the direction of the rotation of the "Chiming Wheel" to clockwise.

5 The rising duck carries a rod on its back that is attached to a vertical rack that turns the cogs.

6 A second rack converts the rotation of the three cogs into a lateral force pushing one arm of the elevator.

7 The "Concertina Elevator" extends, pushing up one end of the lever.

8 As the other end of the lever goes down, the bell is lowered into the path of the "Chiming Wheel."

9 The spinning blades of the "Chiming Wheel" strike the bell, setting off the alarm.

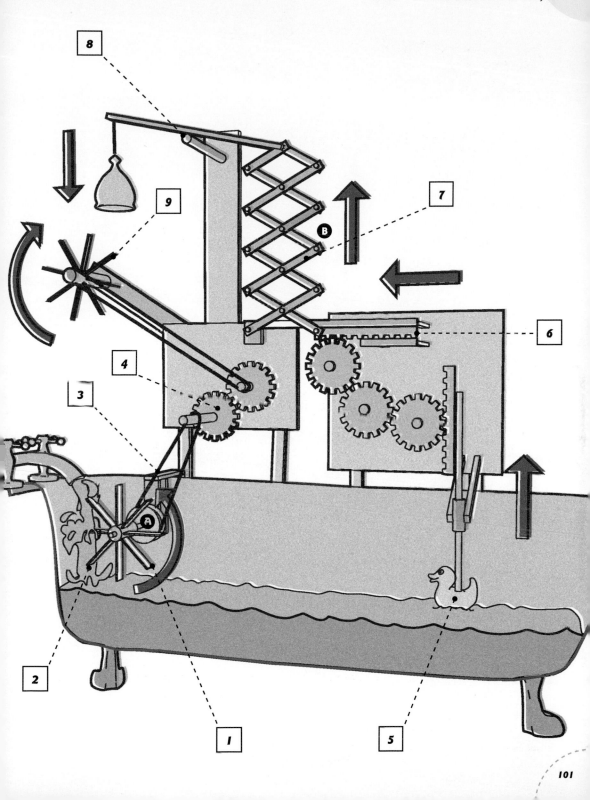

Level: 6

Time: *less than a day*

CONTRAPTION NO.37

The Throne Raiser

This contraption has an entirely different function: raising the toilet seat. The front-end mechanism is the same, but the gears rotate a larger wheel **A** and shaft **B** installed over the toilet. The shaft could be a length of broom handle screwed to the drive wheel through its center. Two counterweights **C** and **D**, which could be large fishing weights, exactly balance the weight of the seat, so that the "Faucet Watermill" can easily provide enough force. The pulley mechanism needs to be firmly attached above the toilet.

String has been wound around the pulleys once to provide traction and prevent the weights from descending prematurely.

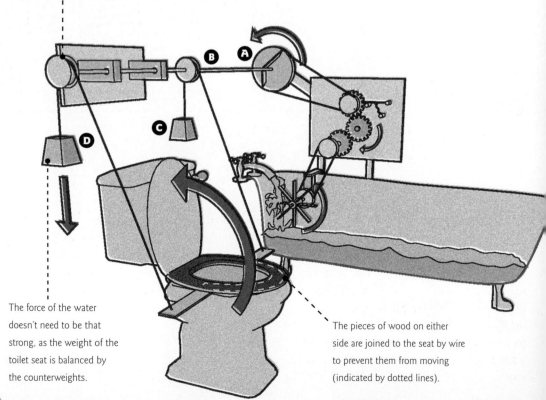

The force of the water doesn't need to be that strong, as the weight of the toilet seat is balanced by the counterweights.

The pieces of wood on either side are joined to the seat by wire to prevent them from moving (indicated by dotted lines).

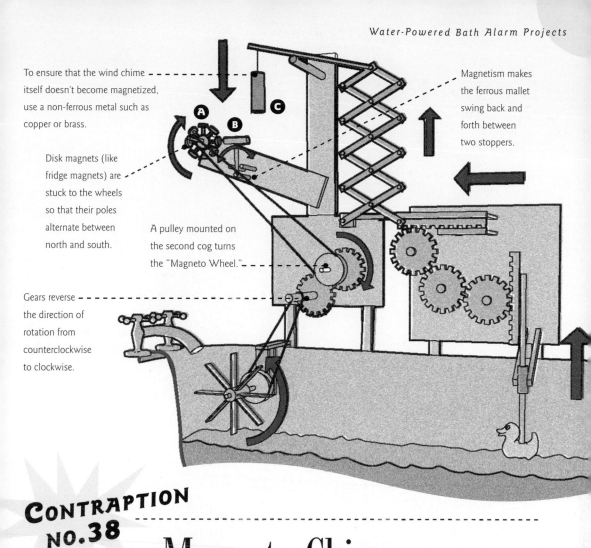

To ensure that the wind chime itself doesn't become magnetized, use a non-ferrous metal such as copper or brass.

Disk magnets (like fridge magnets) are stuck to the wheels so that their poles alternate between north and south.

A pulley mounted on the second cog turns the "Magneto Wheel."

Gears reverse the direction of rotation from counterclockwise to clockwise.

Magnetism makes the ferrous mallet swing back and forth between two stoppers.

CONTRAPTION NO.38
Magneto-Chime

The "Magneto-Chime" is a wheel **A** with an arrangement of magnets set on spokes. Any magnets will do—small disk magnets are usually available in toy stores. The idea is that as the wheel rotates the poles alternate (N, S, N, S). A small iron mallet **B**, set on a pivot, is alternately drawn to and repelled by the magnets as the wheel spins. Because the mallet can't get stuck, it oscillates back and forth. When

the water reaches a certain height, the "Concertina Elevator" lowers a wind chime **C**, made from a length of copper or brass pipe, into the path of the oscillating mallet, where it's eventually struck.

 Level: 8

 Time: one day

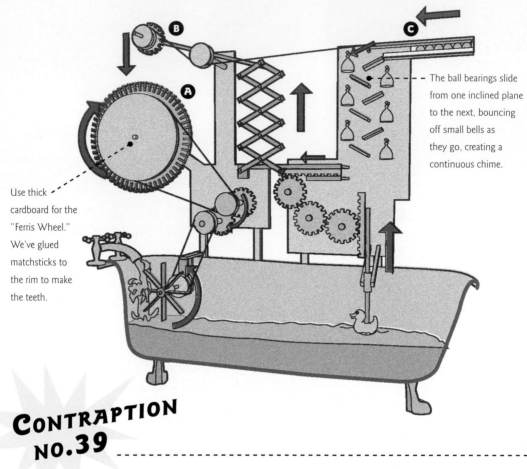

The ball bearings slide from one inclined plane to the next, bouncing off small bells as they go, creating a continuous chime.

Use thick cardboard for the "Ferris Wheel." We've glued matchsticks to the rim to make the teeth.

CONTRAPTION NO.39

The Ferris Wheel

This is one of the more complex projects in the book, but it's worth the effort. In terms of function, we've retained the bath-alarm idea—the power still comes from the watermill and the rising duck—but things get more complicated above the tub. On the left we have the "Ferris Wheel" **A**. You can use matchsticks for the gear teeth, but they'll need to be glued on well to take the strain when they mesh with the teeth on the top cog **B**. When the rising duck causes it to engage, this cog turns and winds in a string that pulls a bar hooked over a row of ball bearings **C**. These travel one by one down the series of inclined planes, ringing the bells as they roll. The top ledge and the inclined planes will need a slightly raised edge to keep the balls on track.

 Level: *9*

 Time: *a day and a half*

CONTRAPTION NO.40

 Level: 6

 Time: half a day or more

Bubble Bath Whisker

For the final project in this section, we've gone for a simplified alarm and an additional function: stirring the bathwater to make sure you get a good head of foam on your bubble bath. The "Faucet Watermill" powers two "Bubble Bath Whisks" **A** that rotate in opposite directions. These can be actual whisks, or you can make them from stiff sheet plastic. The whole assembly pivots on the shaft that carries the watermill **B**, which can be made from a metal rod. As the floating end of the assembly rises, a knob on the top gear wheel strikes the bell, informing you that the bath is full.

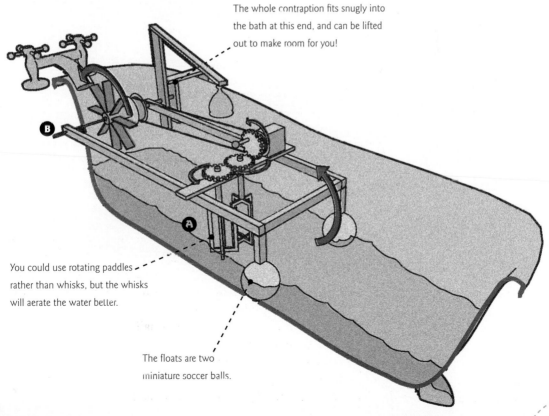

The whole contraption fits snugly into the bath at this end, and can be lifted out to make room for you!

You could use rotating paddles rather than whisks, but the whisks will aerate the water better.

The floats are two miniature soccer balls.

CONTRAPTIONS 41–45

In these energy-conscious, carbon-footprint-neutral, eco-friendly, green-aware days, we all have to think about the future of the planet. Even if you have installed energy-saving light bulbs all around your hearth and home, you still have to remember to switch them off. Here's a group of contraptions that'll help you to remember, if nothing else.

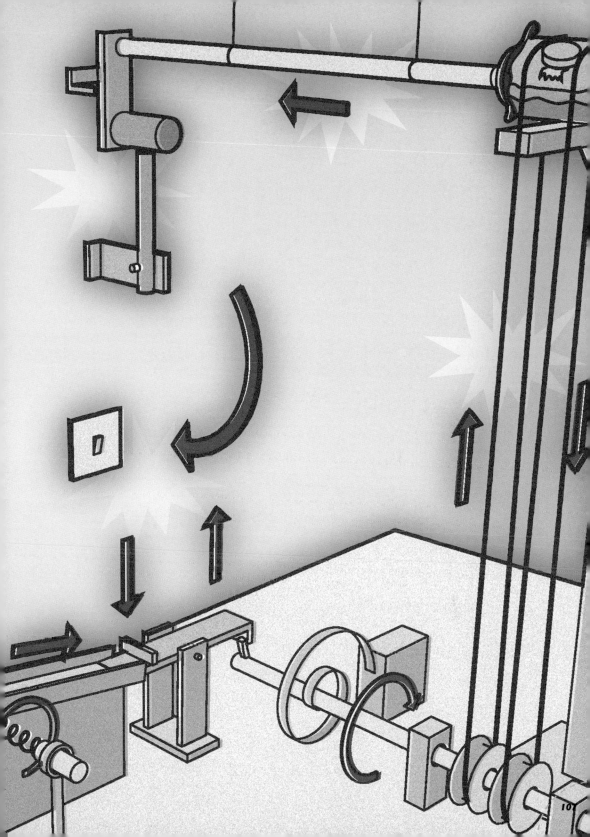

Energy-Saver Light Switch Projects

How many times have you got into bed, only to realize that you left the light on? You can try to ignore it, turn to face the other way, bury your head in the pillows, but somehow that barely perceptible glow is like a thousand-megawatt search beam seeking you out. You have no choice but to get up and turn it off. Or do you? If you'd installed any one of the following clever contraptions, you could drift off untroubled.

Power: *gravity, spring, air pressure*

Principles: *projectile, lever, incline plane, pulley, wheel and axle*

The Inspiration

If you've ever lived in the kind of home with a light switch at the bottom of the stairs but not at the top, or at one end of the corridor but not the other, you'll know why this contraption is particularly useful.

Trouble is this looks like so much fun that you might want to hang around to see it work, and still have to make it to the bedroom in the dark.

Rules and Levels

This pretty much has everything that contraption engineering has to offer: a variety of novel principles and power sources; recycled and improvised parts; ridiculous complexity to accomplish a simple task; and a big finish! Because it's

---

SUGGESTED COMPONENTS

wood • two cotton reels • weight • spiral binding from notebook • cork • suitcase padlock • ball bearing • strip of plastic • glass jar • string • two soluble fizzy tablets (for indigestion or similar) • water • rubber balloon • wooden mallet • and a suitably placed (and solid) light switch...

---

fairly elaborate with its six component devices, we've given it a 7 out of 10, but because each one is not that difficult to put together, we think you should be able to get it up and running in less than a day.

Bits and Pieces

Push the weight off the ledge to get the whole thing going. The weight rotates a rod attached to the wire spiral binding from a notebook or artpad, the other end of which is fixed onto a cork. Hanging onto the spiral binding there is a small padlock of the type you get with a bag or a suitcase—almost a toy one. As the spiral binding rotates, the padlock is drawn along it—because it's a kind of screw. It reaches the ball bearing and knocks it down the slope. The moving ball releases a catch that frees a spring made of a strip of plastic. The rod rotates, and with it the wheels that are linked by string to a small jar (of the type you use for jam or honey) sealed at one end with a piece of rubber. Inside there is water and a couple of fizzy tablets attached to the dry side by a piece of tape. As the string rotates the jar, the tablets dissolve. The "Gas Bottle" activates the "Hammer Striker," which is a small wooden mallet.

Power and principles

That old contraption-engineering favorite, the gravity/pulley arrangement provides the initial motive power, turning the "Helix Padlock Conveyor." The padlock knocks the ball, a projectile, down an inclined plane, at the end of which it trips a first-class lever. This frees the "Spring-Powered Rotator," that turns the pulleys at one end of the rod. The rod is linked to a jar—the "Gas Factory"—by a couple of loops of string. When the rope turns this, it immerses the tablets in water. As they dissolve, they fill the jar with gas, changing the pressure and pushing the rubber membrane that seals the jar outward. The membrane displaces the rod that trips a first-class lever, releasing the mallet—finishing where we began, with gravity.

Variations on a Theme

Contraptions NOS.42–44 (see pp. 112–114) retain the switch function, but mix up the principles and power sources; but Contraption NO.45 (see p. 115) goes for another finishing device: the helicopter.

Hammering it home: these contraptions finish with a bang.

Contraptions NOS.42–44 (see pp. 112–114) ... Contraption NO.45 (see p. 115)

SPIRALING AROUND

Although the binding for a notebook is known as a "spiral" binding, it should be a "helical" binding. A spiral is flat, while a helix is a three-dimensional coil that runs around a central cylinder. The helical (spiral) binding for a notepad is shaped by winding wire around a rod, which is then removed. Helixes are common in the manufactured and natural worlds, the screw being the most common example of the former, and the double helix of DNA the latter.

CONTRAPTION NO.41

 Level: 7

 Time: less than a day

Energy-Saver Light Switch

As well as being fun and challenging to make, this contraption is great to watch. Once set in motion, it combines some slow-and-steady processes with some very fast action, especially the final hammer swing. Some fairly precise alignment is needed to make sure the padlock knocks the ball down the slope, and for the hammer to be finely balanced and accurate when it makes contact with the light switch.

1 We've used a lot of cheese as our initial weight, but any suitably shaped weighty object will do.

2 The helical binding rotates, moving the padlock along its length.

3 The padlock knocks the ball bearing down the inclined plane, tipping a lever that acts as a catch for the spring-loaded rod.

4 The spring is cut from a sheet of stiff flexible plastic. Metal will work, but may be too violent.

5 Twin loops, turned by the pulleys, rotate the jar, which is half-filled with water.

6 A couple of fizzy tablets, which must be dry, are taped to the top of the jar.

7 Wetted in the rotating jar, the tablets dissolve, filling the jar (now the "Gas Factory") with CO_2.

8 The expanding gas pushes out the flexible membrane that seals the open end.

9 The membrane pushes the rod, which tips the lever, releasing the "Hammer Striker." Goodnight!

CONTRAPTION CONNECTIONS

The only ready-made item you might want to buy specially for this project is the wooden mallet **A** (and even this you could easily make yourself). You'll find the remainder of the components easily enough in the average home or garage. If you don't have one of those small toy padlocks **B**, a weighted keyring will work. The supporting structures need to resist the movement of the parts, so use wood rather than cardboard. The membrane that seals the jar **C** can be cut from a bicycle tire's inner tube, or you could use a piece of a balloon.

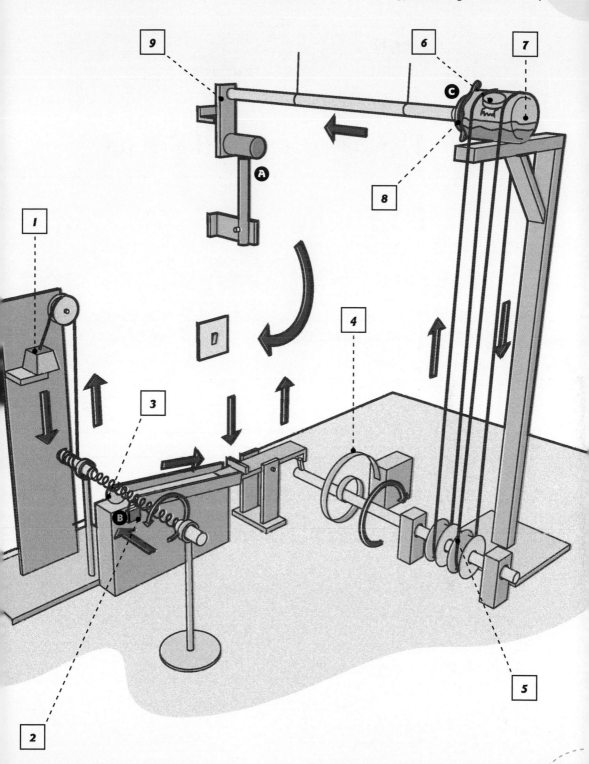

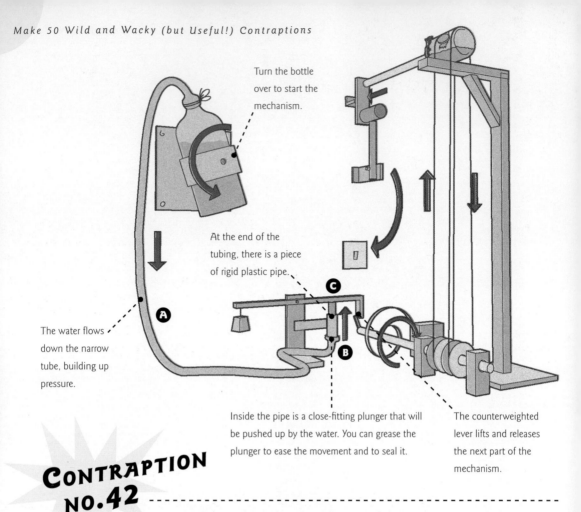

Turn the bottle over to start the mechanism.

At the end of the tubing, there is a piece of rigid plastic pipe.

The water flows down the narrow tube, building up pressure.

A

B

C

Inside the pipe is a close-fitting plunger that will be pushed up by the water. You can grease the plunger to ease the movement and to seal it.

The counterweighted lever lifts and releases the next part of the mechanism.

CONTRAPTION NO.42

The Jet Stream

For this contraption, we return to water pressure as the initial source of power. The pressure builds up in the tube **A** when the bottle is inverted. Rubber tubing has the flexibility you need, and can be stretched to fit the bottle neck. Once it reaches a certain level, the pressure activates the plunger mechanism at the other end of the tube **B**. This can be made from plastic or copper piping, into which a close-fitting piece of dowel slides. It needs to be free to move up smoothly, but without letting the water squirt past it, or there won't be enough pressure to push up the counter-weighted lever **C** and release the spring-loaded rotating rod that sets off the rest of the sequence.

Level: 5

Time: *half a day*

CONTRAPTION NO.43

 Level: 6

 Time: less than a day

The Double Helix

This contraption uses a new mechanism to push the mallet down. We've lengthened the rod and turned it into a second helical screw. You can use a strip of thick cardboard, or even flexible plastic tubing, wound around it to create the helix **A**. Cardboard can be glued, but tubing will need to be wound tightly and nailed at each end. Two slanted pieces of wood or thick cardboard **B** are attached to the supporting sub-structure, and as the rod rotates it literally screws itself forward against them, pushing itself into the lever **C** on which the wooden mallet is balanced. When the rod finally reaches the lever it triggers the hammer and "Crash! Bang! Wallop!" and darkness falls.

As the rod rotates it's screwed forward toward the lever that balances the "Hammer Striker."

We've used a long section of wooden dowelling to make the rod.

Cut a long strip of cardboard and wind it around the rod to create the helix.

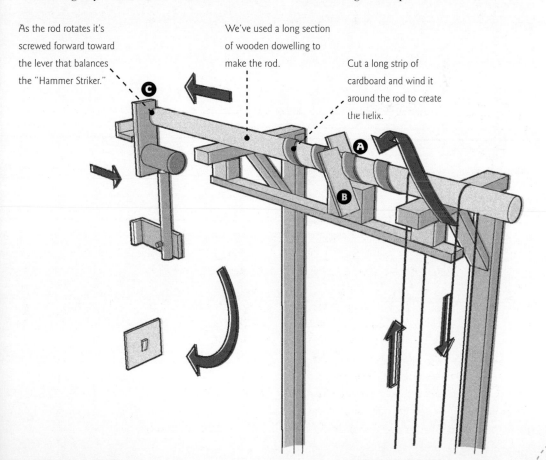

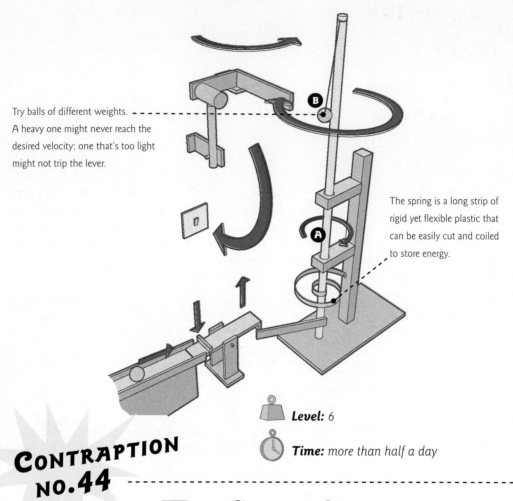

Try balls of different weights. A heavy one might never reach the desired velocity; one that's too light might not trip the lever.

The spring is a long strip of rigid yet flexible plastic that can be easily cut and coiled to store energy.

Level: 6

Time: *more than half a day*

CONTRAPTION NO.44

The Centrifuge

This contraption owes its name to the new component that has been introduced to trigger the mallet. When the rolling ball trips the lever and releases the catch, the spring that goes into action is set at the base of a vertical rod **A** rather than a horizontal one. The rod can be a piece of dowel. At the top is "The Centrifuge" itself. This is basically a ball on the end of a string **B**. The ball can be made of rubber or wood, or you could try a golf ball. As the rod spins, the ball swings in a widening circle, until it collides with the lever and releases the "Hammer Striker." If the ball misses, try changing the length of the string or moving the rod closer to or farther from the lever.

CONTRAPTION NO.45

 Level: 7

 Time: more than a day

The Helicopter Launcher

For this contraption you can either retain the front section, or just start from the point at which the ball rolls down the slope. Two pulleys on one end of a horizontal rod are joined by string to further pulleys **A** and **B**, one at each end of the "Launch Pad." As the rod rotates, the string draws two blocks in opposite directions along a set of guide rails **C**. Squeezed between the blocks is the stem of the helicopter **D**. As the blocks slide, they spin the stem quickly, so that the helicopter shoots into the air. The guide rails can be made of wood or metal, but must be firm enough to keep the blocks pressed against the stem, which is kept upright in a metal tube.

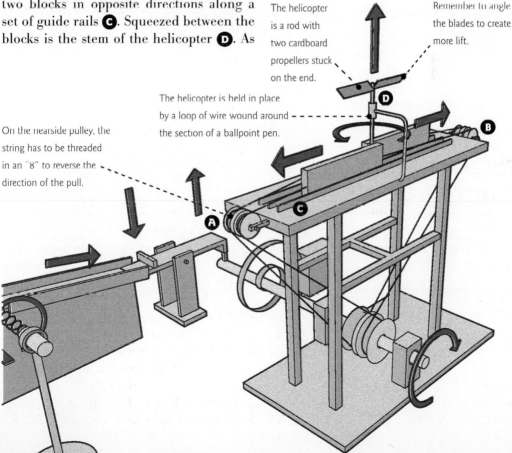

The helicopter is a rod with two cardboard propellers stuck on the end.

Remember to angle the blades to create more lift.

The helicopter is held in place by a loop of wire wound around the section of a ballpoint pen.

On the nearside pulley, the string has to be threaded in an "8" to reverse the direction of the pull.

CONTRAPTIONS 46–50

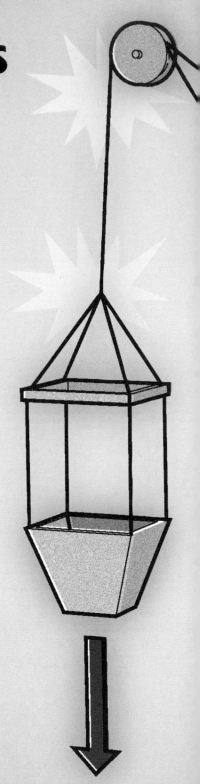

Chances are that not many people have domino sets these days, and that anyone under twenty hasn't ever seen a game of dominos, let alone played one. So these contraptions could also be used to upend any small rectangular object—a matchbox, for example—that had inexplicably and unexpectedly been flipped over.

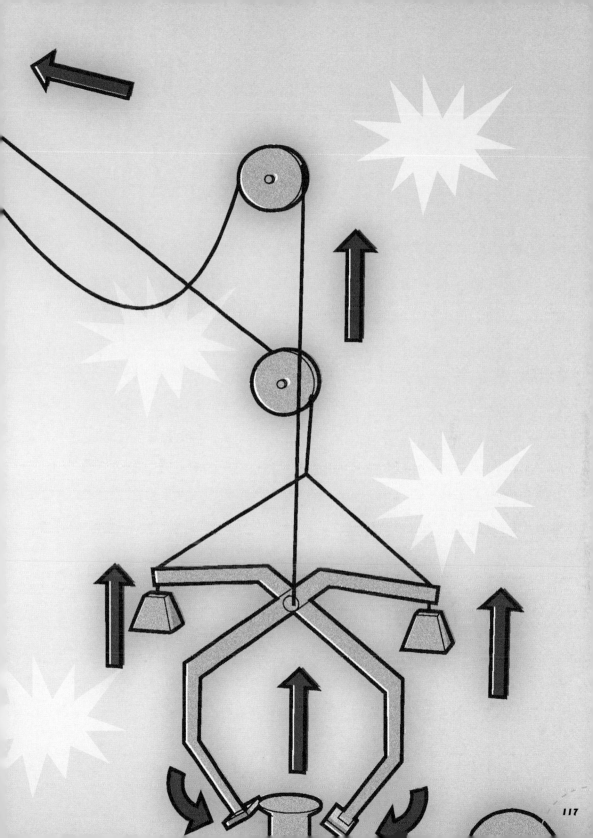

117

Gravity Domino Righter Projects

Dominos was at its most popular at the same time as cartoons of mechanical absurdities became popular; but you don't need to own a domino set to make use of these handy contraptions. Any small rectangular object will do—for example, a matchbox.

Power: *gravity*

Principles: *pulley, lever*

The Inspiration

For anyone not familiar with the game: Dominos is a game of skill in which the winner is the first to discard all his or her dominos, or "bones." A player picks a bone from a pile, and can lay it down on the playing area if he or she can match one of the two numbers inscribed on one side (from 0 to 6, represented as spots, or "pips")

SUGGESTED COMPONENTS
wood • plastic flowerpot • sand • string • three weights • grapefruit • and a domino...

---

with one of the numbers on a bone already in play. Bones that aren't in play are laid face down with their pips hidden, hence the need for a domino-righting machine. However, don't despair, even if you haven't got a domino set, this contraption will flip any small rectangular object.

Rules and Levels

This project and the three that follow are good examples of how you can gradually build complexity into a design by adding extra stages to the contraption. A second principle of contraption engineering that this project illustrates is making a single part serve a dual purpose. In this case, the "Counterweight and Crane," which first grabs the weight, and then lifts it. As the basic design only has three components: the "Gravity Bucket," the "Counterweight and Crane," and the "Softball Seesaw," we've given it a 5 out of 10, with a build-time of half a day. As you'll see in the next three projects, however, things can get a lot more complicated.

Bits and Pieces

Most of this contraption is going to be wood, with the pulleys made of cotton reels, as usual. The counterweights can be made of cheese or small blocks of any suitably heavy material. The device kicks off when you pour sand into the "Gravity Bucket," in this case, a plastic flowerpot (make sure you block up the hole in the bottom). The bucket is connected to a series of pulleys that activate the "Counterweight and Crane" mechanism, which first grabs then lifts the weight, which releases the softball that provides the last push to the "Domino Righter."

Power and Principles

Gravity plays a major role in this contraption, as it's utilized in both the starting and finishing mechanisms. The force it generates is transmitted through a system of pulleys and levers. The grabbing action of the "Counterweight and Crane"

DOUBLING UP

The most common example of a double first-class lever is the humble pair of domestic scissors. Like a single first-class lever, a double first-class lever makes use of the increased mechanical advantage by placing the material to be cut as close to the fulcrum as possible. For example, heavy-duty bolt cutters have long handles when compared to the short cutting head. In addition to leverage, scissors make use of the shearing force of the two blades.

uses the principle of leverage. In this instance a double first-class lever with the central counterweighted pivot acting as the fulcrum (see box above).

Variations on a Theme

For Contraptions NO.47 (see p. 122), NO.48 (see p. 123), and NO.49 (see p. 124), we've retained the domino-righting formula, but added complexity to each successive project, but for NO.50 (p. 125) we've gone for our version of a contraptioneering classic: the "Humane Mousetrap."

These "Domino Righter" contraptions use a rolling ball to give that domino the nudge it needs.

UPLIFT

Cranes have been in use since antiquity, when they were in use by the Greeks and Romans, famous builders of huge temples, city walls, and aqueducts. Medieval cranes were used in construction, as well as to load and unload merchant ships. The crane makes use of a couple of simple machines (see pp. 16–17) to increase its overall mechanical advantage: The horizontal lifting arm is a lever, and the hoisting mechanism includes one or more pulley arrangements.

CONTRAPTION NO.46

Level: 5

Time: half a day

Gravity Domino Righter

For the proficient contraption engineer that you should have become by this project, the "Gravity Domino Righter" will not present any major challenges, with the possible exception of getting the timing right for the grab-and-lift action of the "Counterweight and Crane" mechanism. The amount of slack that you leave in the second string will determine when the lifting begins, and this should be as soon as the crane has fully grabbed the weight. The system of weights and pulleys and the lever and rolling ball are straightforward.

CONTRAPTION CONNECTIONS

The crane's jaw mechanism **A** could be made of wood, but because of the complex shape it's more easily cut from thick cardboard. This keeps it light, and the counterweights **B** and **C** can then be small fishing weights, just heavy enough to hold the jaws apart. The weight being lifted **D** needs to have a lip around the top edge below which the jaws can catch. A heavy cotton reel is ideal, but it must be heavier than the ball, unless you move the fulcrum closer to the ball to compensate. The two levers need to be stiff, so wood is better than cardboard for this purpose.

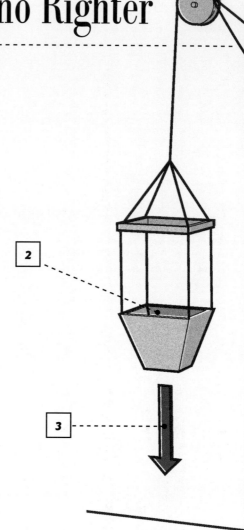

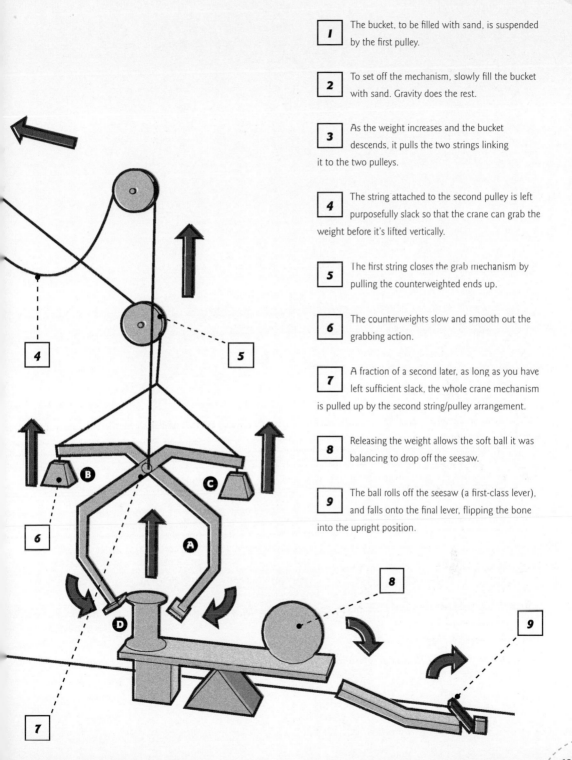

1 The bucket, to be filled with sand, is suspended by the first pulley.

2 To set off the mechanism, slowly fill the bucket with sand. Gravity does the rest.

3 As the weight increases and the bucket descends, it pulls the two strings linking it to the two pulleys.

4 The string attached to the second pulley is left purposefully slack so that the crane can grab the weight before it's lifted vertically.

5 The first string closes the grab mechanism by pulling the counterweighted ends up.

6 The counterweights slow and smooth out the grabbing action.

7 A fraction of a second later, as long as you have left sufficient slack, the whole crane mechanism is pulled up by the second string/pulley arrangement.

8 Releasing the weight allows the soft ball it was balancing to drop off the seesaw.

9 The ball rolls off the seesaw (a first-class lever), and falls onto the final lever, flipping the bone into the upright position.

A simple switch that you push down releases
a rubber ball tied to the end of a string...

...and it releases the pulley and
weight assembly that powers the
grab mechanism of the crane.

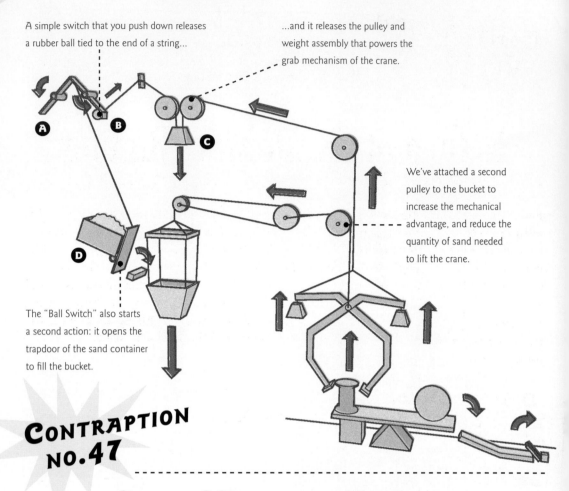

We've attached a second
pulley to the bucket to
increase the mechanical
advantage, and reduce the
quantity of sand needed
to lift the crane.

The "Ball Switch" also starts
a second action: it opens the
trapdoor of the sand container
to fill the bucket.

CONTRAPTION NO.47

Son of Domino Righter

This project adds a second gravity device, which actually slightly simplifies the mechanism by supplying separate power sources for the counterweight and the crane. The new element starts with a switch **A** (perhaps made of wood, and pivoting on dowels) that releases a ball **B** on the end of a string. The released ball allows a weight **C** to descend and close the grab. The initial switch also triggers the opening of a trapdoor on the container **D**, on the bottom left, that fills the bucket with sand. The descending bucket powers the crane's lifting action, and from there the rest of the contraption works as shown.

Level: 6

Time: *more than half a day*

CONTRAPTION NO.48

Level: 8

Time: one day

Return of the Domino Righter

For advanced contraptioneers, we've added even more complexity to this project. The new starting device, the "Roller of Death," consists of a cylinder (a piece of thick dowel or section of broom handle works well)

fitted with four ball bearing counterweights on arms **A** to keep it from falling off the five inclined wooden slats **B** as it picks up speed. The revolving arms first trip the switch **C** to start the grab mechanism, and when the roller reaches the bottom of the run it strikes the catch **D** that holds the trapdoor of the sandbox closed.

The counterweighted arms on the roller keep it on the straight and narrow path.

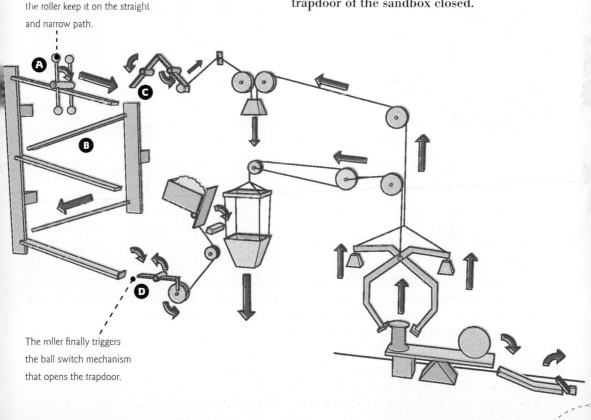

The roller finally triggers the ball switch mechanism that opens the trapdoor.

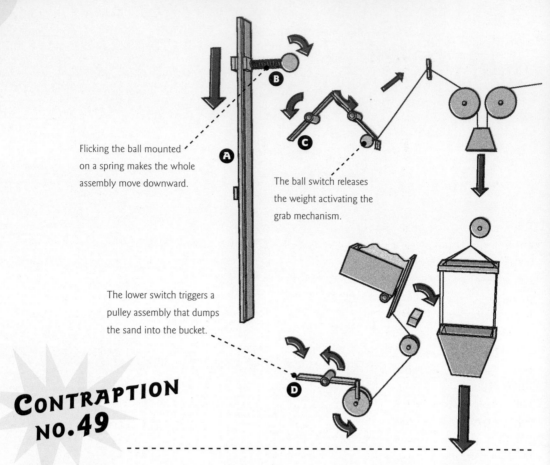

Flicking the ball mounted on a spring makes the whole assembly move downward.

The ball switch releases the weight activating the grab mechanism.

The lower switch triggers a pulley assembly that dumps the sand into the bucket.

The Domino Righter Strikes Back

Do you remember the toy woodpecker that staggers its way down a pole? A flick of its tail vibrates a spring that sets the bird pecking at the pole as it goes down. The new starting mechanism for this device is a simplified version of the woodpecker toy. Two strips of wood **A** hold a block between them while still allowing some movement. A rubber ball on the end of a spring **B** is fixed to the block. When you flick the ball, making it oscillate on the spring, the block will slide down slowly in staggered steps. The descending block and ball trigger the first switch **C** to power the crane grab, and then, just a moment later, the second switch **D** activating the lifting mechanism. The ball should be made of solid rubber, to give it the necessary weight, and the spring should be fairly stiff to impart the motion to the block.

 Level: 7

 Time: more than half a day

CONTRAPTION NO.50

Level: 8

Time: *more than a day*

Humane Mousetrap

Where better to finish a book on contraption engineering than with a "Humane Mouse-trap"? The front end of the machine can be set up as in any of the previous projects, so pick whichever one you want to use. But the business end is completely new. As soon as the mouse climbs one of the sloping ramps **A** and **B** and stops on the bridge to investigate the tasty morsel that you have left out for it, you set the contraption in motion. As the unsuspecting mouse is sitting happily munching, thinking mouse thoughts, the first string pulls up the hinged drawbridge mechanism **C** on which it is standing. The surprised mouse falls into the box (a tiny mattress will break its fall) just as the second string shuts the door **D**. All of these elements can be made from stiff cardboard, unless you're planning on catching an exceptionally heavy mouse.

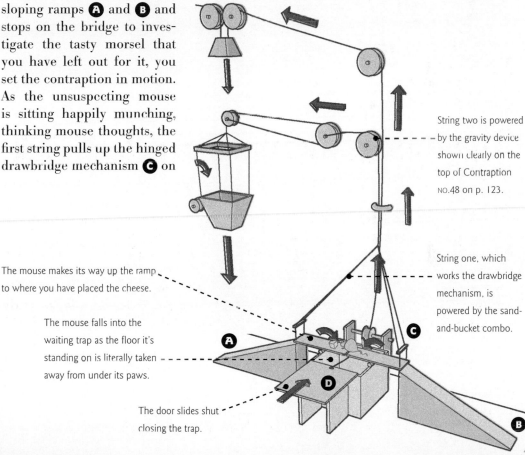

String two is powered by the gravity device shown clearly on the top of Contraption NO.48 on p. 123.

The mouse makes its way up the ramp to where you have placed the cheese.

The mouse falls into the waiting trap as the floor it's standing on is literally taken away from under its paws.

String one, which works the drawbridge mechanism, is powered by the sand-and-bucket combo.

The door slides shut closing the trap.

INDEX

Resources

All of the contraptions in this book have been designed so that you can make them from whatever materials (and to whatever scale, large or small) you want. Each component—whether it's a lever, cog, or pulley—can be made from common household items, or from materials just as easily found at your local craft or hardware shop. For some suggestions on tools and supplies, see pp. 14–15; for purchasing items, here are some handy online resources.

Ace

www.acehardware.com

Ace is a nationwide chain of hardware stores selling a vast range of tools and materials. Products are available online and in-store.

Create for Less

www.createforless.com

Create for Less sells more than 50,000 brand-name craft supplies at wholesale prices (and in bulk, if you feel like building lots of contraptions).

Do It Best

www.doitbest.com

An online store representing over 4,100 independently owned hardware and home-improvements stores, Do It Best offers over 65,000 items.

Home Depot

www.homedepot.com

An exhaustive resource for tools and hardware, Home Depot's motto is "You can do it. We can help." So let them!

Lowe's

www.lowes.com

The second largest home-improvement retailer in the world, Lowe's offers the building products, tools, and accessories for making just about anything.

Michaels Stores

www.michaels.com

The largest specialty retailer of arts and crafts materials in the US, with over 900 stores located throughout North America.